Untangled web:
developing teaching
on the internet

Pearson
Education

We work with leading authors to develop the
strongest educational materials in technology and
education, bringing cutting-edge thinking and best
learning practice to a global market.

Under a range of well-known imprints, including
Prentice Hall, we craft high quality print and
electronic publications which help readers to
understand and apply their content,
whether studying or at work.

To find out more about the complete range of our
publishing please visit us on the World Wide Web at:
www.pearsoneduc.com

Untangled web:

Untangled web:
developing teaching on the internet

David T Graham
Jane McNeil
Lloyd Pettiford

Prentice Hall

An imprint of **Pearson Education**

Harlow, England · London · New York · Reading, Massachusetts · San Francisco · Toronto · Don Mills, Ontario · Sydney
Tokyo · Singapore · Hong Kong · Seoul · Taipei · Cape Town · Madrid · Mexico City · Amsterdam · Munich · Paris · Milan

Pearson Education Limited
Edinburgh Gate
Harlow
Essex CM20 2JE
England

and Associated Companies around the world

Visit us on the World Wide Web at:
www.pearsoneduc.com

First published 2000

© Pearson Education Limited 2000

ISBN 0 582 41854 2

British Library Cataloguing-in-Publication Data
A catalogue record for this book can be obtained from the British Library.

Library of Congress Cataloging-in-Publication Data
A catalog record for this book can be obtained from the Library of Congress.

10 9 8 7 6 5 4 3 2 1
04 03 02 01 00

Typeset by 35
Printed in Great Britain by Henry Ling Ltd., at the Dorset Press, Dorchester, Dorset

For Maureen, Jon and Norksy,

and not forgetting Anne and Bill

Contents

Figures

Tables

Acknowledgements

We would like to thank The Nottingham Trent University which provided us with a grant from its Teaching and Learning Development Fund. This allowed us to develop the projects discussed herein and also helped meet some of the costs of producing this book. We must thank Linda Dawes of Belvoir Cartographics and Design for a first class job typesetting the book. Jonathan McLachlan offered technical advice and proofreading. Maureen Docherty helped with proofreading and referencing. Finally, thanks are due to Matthew Smith of Pearson Education for having faith in the project.

Abbreviations

The following abbreviations have been used in this book:

CGI	Common Gateway Interface
FAQ	Frequently asked question
GIF	Graphics Interchange Format
HTML	Hypertext mark-up language
IR	International relations
ISP	Internet service provider
IT	Information technology
JPEG	Joint Photographic Experts Group
ODL	Open and distance learning
URL	Uniform (Universal) resource locator
WBI	Web-based instruction
WYSIWYG	What You See Is What You Get

A note on web references

The authors have cited many online articles or useful web resources, quoting the relevant URL. As web addresses often change when sites are reorganized, you may find that some of the URLs have expired and no longer lead to a page. If you experience this, try deleting successive parts of the address from the end and this may return to a menu higher up the structure of the site. You can then look out for the article or section cited. Alternatively, use only the root of the address (the part before the first 'forward slash'). This will, in all probability, find the home page of the site, from where you can look for, or conduct a keyword search for, the relevant information.

introduction

introduction

'Begin at the beginning' the King said, gravely,
'and go on till you come to the end: then stop.'

Lewis Carroll *Alice's Adventures in Wonderland* (1865)

Untangled web: developing teaching on the internet

The use of the web in teaching and learning within higher education is now commonplace (Whittington and Campbell 1998). For many teachers using the web in their courses simply means converting their handouts and booklists into HTML, making use of conferencing facilities or providing 'hotlists' of relevant sites. This book shows how educators can go further than this and use the web for delivering stand-alone or integrated teaching packages into their courses. We draw upon our own experiences developing our own web-based instruction (WBI) packages. Khan (1997a: 5) defines WBI 'as an innovative approach for delivering instruction to a remote audience, using the Web as the medium'. We use WBI throughout this book by way of shorthand but would expand Khan's definition and argue that our web-based products go further than simply delivering instruction.

Before we proceed a couple of questions need to be addressed. First, what does it mean to talk of the 'challenge of the educated web?' (Carty 1998). Second, in such a context, why is this particular book needed/how can it help you? In answer to the first question, there is now almost a generalized belief that technology can make academia more accessible, affordable and effective but that this will need changes, probably quite profound, in the use of teaching time and other resources, as well as in the roles of staff and even in the institutional mission itself (Hill 1997; Van Dusen 1998). At the so-called 'chalk-face', models of teaching will have to be changed away from a transmission model to one which is 'much more complex, interactive and evolving' (Sherry 1996: 342). For many teachers this will not be a comfortable change and they will require support. In answer to the second question, there are clearly lots of choices to be made in answering the challenge of the first question. There are many questions to ask and considerations to be made, which amount to cost benefit analyses (Freedman

1995; Potashnik and Capper 1998). However, thus far, it could well be argued that 'internet-based courses usually take more time, expertise and money than expected and deliver somewhat less than is intended' (Levin 1998) and that to date costs have outweighed benefits. As other authors put it, 'the problem with technology driven innovations is that they can consume prodigious amounts of time and money to little educational effect' (Ward and Newlands 1998: 171). Initially, we are attempting to make web-based teaching possible for the vast majority.

However, it is necessary to go beyond such ambitions. Most applications of the web so far have been somewhat crude through lack of reflection on the principles of learning and/or lack of skills in terms of using hypermedia and multimedia (Maddux 1996). The idea of 'instruction' has been defined by Ritchie and Hoffman (1997: 135) as 'purposeful interaction to increase learners' knowledge or skills in specific, pre-determined ways'. In this context they argue that publishing a web page with links to other digital resources does not, in and of itself, constitute the essence of teaching. However, those things that do—motivating the learner, specifying what is to be learned, prompting the learner to recall or apply previous knowledge, providing new information, offering guidance and feedback, testing comprehension and supplying enrichment (Dick and Reiser 1989: 135)—can 'with forethought…be incorporated in instruction designed for delivery on the World Wide Web'. This book is needed, then, to help actualize some of the web's potential in teaching. Since 'progressive teachers who are the early adapters of technology can become change agents for their peers' (Sherry 1996) it is hoped that a snowball effect can be set up.

However, there is a further important reason. On the topic of open and distance learning (ODL) on the web, it has become quite common for people who have done it to write short articles for other (knowledgeable/technically literate) people describing how they have used the web for open and distance learning and what the results have been (Barker 1998a). Such articles assume technical knowledge and, though they describe quite often encouraging or exciting outcomes (Pohan and Mathison 1998), they often leave it to the reader to get this knowledge themselves if they want to follow suit (Tuathail and McCormack 1998). What this book is seeking is the describing and prescribing of a route through the whole developmental process of setting up ODL web pages. In fact, to use this book you do not even need to have the experience of setting up your own web page, though it may require more time and patience for those with only a rudimentary knowledge of IT and web-based technologies. In other words, we do not just discuss why the web is useful for your teaching (see Leu and Leu 1999) but how to make it so. The value of such a work has been recognized elsewhere, for instance in the setting up of web pages produced at Aberdeen University; despite the use of various guides to writing for the web, a great deal of time was spent trying to discover exactly what to do in different situations (Ward and Newlands 1998).

Before going any further, a brief examination of terms will be useful. Open and distance learning implies that the learner will be able to receive and access information and guidance without geographical proximity and that access to this type of learning will be, in some sense, wide. In fact, the technological requirements of some types of ODL have led people to fear that learning may become less open in the sense of restricted to fewer people (Kearsley 1998; Yeomans 1998). On the other hand, openness can be defined more in terms of time and therefore access to learning resources. Though not unconcerned by the implications of anything that might be seen to make learning more elitist, here we deal with ODL mainly in terms of increased access, spatially and temporally. There are additionally a whole host of terms which crop up in seeking to describe new pedagogical approaches informed by technology of some kind of another. Hence terms like resource-based learning, electronic course delivery, net-based learning and so on. Rather than attempt detailed definitions of such terms, this book seeks to explain things carefully, and where confusion is possible, include terms in a glossary.

Though you need not have done so, setting up a web page is now quite a common thing to do for individuals, university courses, companies and so on. The internet is vast; a 'global, publicly accessible communication network that connects together numerous people, groups and organizations in many different countries' (Barker 1998a: 4). However, one side effect of this proliferating activity is that there is now a huge amount of material available on the internet. Furthermore, the growth of information has, to a certain extent, taken people by surprise (Maddux *et al.* 1999). Though the internet was initially heralded as a solution to all sorts of learning and teaching problems, the potentials and problems of such a large amount of information are acknowledged, at least implicitly, by an increasing number of academics (Graham, McNeil and Pettiford 2000). One particular problem is that the quality of existing pages varies enormously (Boshier *et al.* 1997). This huge quantity and range of materials must be tailored and contextualized if it is to support teaching effectively (Shipman *et al.* 1997). This book is, at one level, suggesting a solution to these problems of quantity and quality by helping academics to bring some order to the chaotic world of the web through a series of interlinked and selected web pages which

Untangled web: developing teaching on the internet

will enhance the student learning experience and staff teaching experience.

This is a guide which takes you through various stages of preparation in order to incorporate into your teaching not just a web page, but a fully integrated, interactive, multimedia package with the potential for (organic) growth. This approach seeks to improve upon, though does not supplement, the conversational model of learning which is, the authors would suggest, what universities should be aiming at. In effect, this book makes a contribution towards the notion of supported open learning (Daniel 1999).

If this mix of techno-pedagogy jargon is already too much, it should be re-emphasized that you can use this book in a practical manner as a sympathetic guide through this. It will achieve its aims in a way which will make a web-based approach possible for individual academics, courses or departments with quite limited skills and knowledge in the IT realm. Furthermore, whilst the time and money saving aspects of the web have been overstated, this can be done without a huge increase in resources (Kapur and Stillman 1997). All too frequently, technology-based projects end up mystifying any prospective audience. However, for those requiring a basic guide to using the internet, or who would like one to use alongside this work, a good place to start is Stein's (1999) *Learning, Teaching and Researching on the Internet: A Practical Guide for Social Scientists*. Furthermore, there is a whole range of literature related to the simple use of the internet in learning and teaching (for example, Brooks 1997). Apart from some traditional, conservative attitudes to university pedagogy, the lack of institutional and instructional computer support is, we believe, discouraging attempts to utilize the more systematic use of the web in teaching because of difficulty and frustration. Many universities are failing to get the message regarding incorporation of technology (Daniel 1996) and many staff are left on their own. Whilst others have reported briefly on successful attempts to use the web in ODL projects (see also Riley 1998), this book is a practical attempt to circumvent some of the problems which are frequently associated with institutional inertia, either intellectually or financially based, so that this form of teaching might become a much more frequent occurrence.

Having said that, this book takes you through a full developmental process, although, in another sense, it is likely to be simply a starting point. As we suggest throughout, your web pages, once constructed, have tremendous potential for growth. Such growth may be quantitative and/or qualitative, is likely to be fundamentally organic and may be as much student-led/inspired as anything else. Furthermore, technological advances mean more improvements are likely not just for materials but in terms of students engaged in working on such pages being able to see and hear each other as well as exchanging text messages and working on shared resources (Collis and Smith 1997). In this sense, though web pages may be designed to support a traditional model of university instruction, the web offers the chance of opening up the interaction of student and knowledge in exciting new ways (Duchastel 1997) and integrating skills teaching in a way which is not obviously forced.

Roberts (1999) suggests several interrelated factors in explaining the success of WBI; crucially these include the lecturer's willingness to change his or her methodology and the provision of well-structured learning experiences. However, given the number of variables involved in terms of learning possibilities and outcomes, this book resists the tendency to say 'this is exactly how things should be done'. The book looks simply at possibilities; it does make some strong suggestions, based on experience, but its aim is not to satisfy your hunger by giving you the fish, to use the Chinese proverb recycled by the Commonwealth of Learning, but to teach you how to catch your own fish (Daniel 1999). As Open and Distance Learning materials are developed, there are likely to be an enormous number of variables at play; technology may make the idea of individual academics recede, to be replaced by teams and division of labour. Significantly, there are likely to be cultural differences in how it is thought best to incorporate new ways of teaching which will highlight skills, as much as knowledge and understanding. Faced by the possibility of huge globally provided courses, the provision of locally inspired web resources is perhaps a crucial counter-weight to homogenizing forces in education and society more generally. It should not be forgotten that a successful programme in one location is not necessarily universal and may be less successful elsewhere (Sherry 1996). Whatever the situation, the quality of the pages which fit into whichever scheme will be of paramount importance. Given such a scenario, those who have been 'ahead of the game' will be best placed to view how they fit most comfortably into any new working practices and feel empowered accordingly.

Setting the scene

In quite a short space of time, higher education moved from the 'chalk and talk' era to one in which lecture theatres were kitted out with all manner of audio-visual equipment, and now to an age when the presence of computer systems is having a big impact. According to Barker (1998b), this has happened as a result of the conjunction of five interrelated factors:

- Availability of cheap, easy to use electronic storage facilities.
- Availability of interactive information retrieval software.
- Facilitation of information sharing by global communication networks.
- Relatively easy assimilation of information through multimedia techniques.
- Easy individual/group communication by electronic means.

The web is an extraordinary 'free' source of information which has mushroomed in popularity from limited beginnings in US schools to a global network, presenting numerous challenges for other forms of information delivery (Flake 1996). It is to be hoped that teachers have more success than the French Minitel system in responding to these challenges and ensuring that technology works for and with them rather than against their interests. The convergence of computing technology and telecommunications allow the integration of graphics, text, moving images and communication on a single screen. Technology allows students to go well beyond prescribed texts and almost invites exploration and a pioneering/active approach to learning (Greene 1999) which a library seems, unfortunately, to inspire in all too few students.

In terms of internet-based resources there are a number of online directories to point you in the right direction (see Newton *et al.* 1998). As well as information, the web is a useful way into many IT skills. The insight that multimedia-based open and distance learning is especially suited to match the training needs of a modern information society led the European Union to start and fund the DELTA programme on the subject (Friedrich 1997) and there are many other examples of work in this area (see Khan 1997b). Although higher education is still predominantly associated with shared residence and printed media, some think that open and distance learning, especially using the web, will become the dominant themes in higher education of the future (Nyiri 1997).

In recent years there has been a growing interest in producing course materials on the world wide web. The world market for technology-based learning which was already US$6 thousand million in 1997 is estimated, conservatively, to reach US$26 thousand million by 2005 (Canadian Telework Association 1998). A web-based ODL has several advantages. Material can be accessed from a distance, a wide range of media can be employed, conferencing can be undertaken and so on. Computer networks open up new possibilities for the support of synchronous and asynchronous learning/teaching activities (Barker 1998a; Berge 1999; Shotsberger 1997). With IT skills now regarded as key skills, such courses can provide an important and interesting way to introduce them into the university curriculum.

However, there are problems in setting up and using ODL packages, most notably in terms of time and skills on the part of the teacher or teaching team. There are additional problems with ready-made packages (Chen and Zhao 1997) and these are discussed in Chapter 1. This book draws on the experience of the authors who have successfully used web-based projects as integral parts of level three, two and one modules in the Department of International Studies at The Nottingham Trent University (as well as other smaller projects) in order to suggest what works and what does not. We do not deny that further research into pedagogical issues surrounding the web is highly desirable (Windschitl 1998); on the other hand, progress will undoubtedly be made through application rather than purely theoretical reasoning and proceeding on the basis of 'what works' will be a necessary supplement to pedagogical/theoretical reflection.

This book addresses the issues related to the successful integration of specifically tailored, web-delivered material as an integral part of existing, traditionally taught modules. It also examines some of the practical issues surrounding WBI in the light of attempts to reconcile the costs of such innovation with the educational benefits (Ruppert 1998). An evaluation of this type of delivery as an enhancement of the undergraduate experience is discussed in terms of open and distance learning. Since this is, in essence, a practical guide, this introduction and Chapter 1 seek primarily to contextualize our efforts and come complete with a range of references for those who want to read around the history of such projects and the problems and potentials encountered by other projects.

The writing team come from a range of backgrounds. Jane McNeil teaches medieval history and has a long-term interest in the internet. David Graham teaches geography and, although possessing a long-term interest in information technology (IT), is a relative newcomer to the web. Lloyd Pettiford teaches international relations and is something of a novice in terms of IT and a neophyte to the web. We hope that this range of backgrounds, variety of subject areas and different approaches to using the web in our teaching will stimulate and encourage others with aspirations to use the web for teaching or make more use of the web in their teaching. We have each developed our own WBI, based on modules we teach at The Nottingham Trent University.

The original project was developed in 1996 as a supporting element to a social geography module taught to second year students. We wanted to use the media as a way of expanding the horizons of the undergraduates but also to use the web as a means of broadening access through open and distance learning (ODL). By embedding and combining media within web pages the basic but effective *Social Geography and Nottingham* (Mark I) WBI was born, building on the material in the module but using local examples with which the students would have some familiarity. This proved to be a great success with the students. It was decided to expand and develop this into a Mark II version. This has been test driven and approved by a further cohort. It is now in its third incarnation and still going strong.

Such was the success of this project that we decided to expand the scope of web-based teaching. We were successful in bidding for internal funding which allowed us to purchase equipment and software and employ an extra pair of hands in the shape of a research assistant. Thus a new WBI dedicated to a third year environment in international relations module was developed and *e:net* was born. This had different aims from *Social Geography and Nottingham* and was targeted at a completely different audience. Once again, this proved to be a popular addition to an existing module.

The baby of the family is the *Medieval History* WBI which supports the first year medieval history module. This is a team-taught module and has been set up as a study aid. Early indications suggest that this site is popular with the class.

This book is based on our experiences of devising, designing, constructing, testing, adapting and evaluating three fully functional web-based teaching and learning products. These were all written from scratch and based on HTML. Although some WBI authoring software is available (Goldberg 1997; Hansen and Frick

1997), maximum flexibility is achieved by using the hypertext that is the backbone of the web.

The web can be used in a number of ways to enhance education and deliver material. A number of models have been developed for a variety of pedagogical purposes. These are:

1. Information-based models — the web is used for retrieving information, as in virtual museums and digital libraries.
2. Teaching, media-based models — the web is used only for dissemination of educational material to off-campus students in the form of module handbooks, booklists, software, and the like.
3. Enriched classroom models — ODL techniques are used via the web to complement traditional classroom-based teaching.
4. Virtual classroom models — the web is used with emphasis on collaboration and computer mediated interaction (Retalis *et al.* 1998: 16).

Our WBI projects cover some of these, especially models two and three.

How to use this book

Just as we stress throughout that WBI is a non-linear process, so your approach to this book should be on a need to know basis. Do not feel that you have to heed the advice the King gave to Alice. You will probably wish to read this introduction as a preliminary. However, though the chapters are written in sequence, there is no requirement to follow our ordering rigidly. Some users might feel confident with the material in some chapters and choose to skip these. This is simply the best order that we have found to work to.

It is definitely not designed as a manual. Many useful and not so useful books and articles are available about the internet, the web, web design, HTML and the like. Nor is this a book about ODL or using new technologies in teaching. Many good books and articles on open and distance learning exist and are referenced in this work. What this book offers is our combined experiences of constructing WBI from scratch.

Chapter 1 begins by considering the pedagogic logic of using the web as part of your teaching. It starts with conscious reflection on what your needs are. It is important not to visualize web pages as simple tack-ons to learning, but as fully integrated additions. Teaching has incorporated the internet in recent times in this way, but this is felt to be unsatisfactory for a number of reasons. Simply providing information in a substitutive fashion adds little to the learning experience. Thus the issue of integrating web-based projects is crucial. Also in Chapter 1, the issue of evolving and quantifying learning outcomes is addressed. This is an exciting time with much research into pedagogic issues and web-based packages taking place (Edwards 1996). Chapter 1 might be said to stem from and develop Barker's (1998a) 'four major ongoing transitions'. These can be summarized as, first, a move from instructivist learning and teaching philosophies towards ones based on principles of constructivism. Second, a shift away from tutor control. This means moving to a situation where students are responsible for specifying and managing their own learning activities. As flexibility in terms of career progression becomes increasingly salient to the graduate employee, these are skills which will be ever more valuable. Third, there is a move to a situation where learning materials are much more universally available. That is, they are not specific to a particular time and place but are available at any hour and from any location. Finally, computers are increasingly being used in what Barker (1998a: 4) terms 'computer mediation'. By this he means the use of computers to back up, reinforce or extend knowledge transfer leading to the 'development of rich mental models'. Numerous individual case studies attest to the numerous benefits of using the web to facilitate active learning (Yaverbaum and Liebowitz 1998).

Moving to Chapter 2, the need for this book stems from one particularly important reason. Since the web is exciting and new, in a rush to use new technologies, many people have tended to overlook the design aspects of online hypertext (Barron *et al.* 1996; Falk 1997). In Chapter 2, issues of planning the information architecture are addressed—concerning content, functions, structure and thought paths. Architecture is the key notion. A house would not be considered very desirable or sensibly designed if all the rooms were ranged in a straight line with the only way from one to the next being a single connecting door. Similarly if all were kitchens! Much the same can be said of an interconnected set of learning resources; a linear approach may not be most suitable, though as with traditional teaching it may sometimes be necessary. Chapter 2 thus discusses a range of issues designed to ensure a desirable residence for your teaching materials, since the nicest of furniture can look out of place in inappropriate surroundings. Furthermore, you have to make people want to stay in your house. Though an advantage of WBI is the freedom students have to pursue their own thing, this can lead to a superficial approach. Your design needs to make people want to stay, and you, therefore, need to encourage students to do something with the knowledge they are presented with. Your information, however good it is, must be organized and sensibly arranged (housed) if it is to prove a useful learning resource (Schwier and Misanchuk 1996).

Chapter 3 takes the above analogy a stage further. Designing a successful house is not just a question of deciding which rooms go where but what the place looks like, how easy it is to live in and how to get around. Will your students be able to do this, even if they do not enter from your main page/orientation section? There are then certain practical, navigational, considerations; a house with too much state of the art technology might be difficult to understand or expensive; maybe even difficult to get around. Similarly, in designing web pages there are various technical constraints to be addressed and trade-offs between what is desirable and what is practical. The downfall of many individual/ private web pages is an inability to sacrifice an all singing, all dancing approach for the benefit of those who might visit the site.

Having decided against linear arrangement of rooms and a jacuzzi in the kitchen, the next stop is furniture. What material will you use in your pages? Clearly this is related, to a large extent, to the particular subject you are teaching. However, there are important issues related to copyright (Douvains 1997), utilizing multimedia and assessment, which can affect content, and these are discussed in a general sense in Chapter 4. Part of the wasted time referred to by Ward and Newlands (1998: 174) above in talking about setting up web pages was due to 'uncertainty about the law of copyright relating to publishing in electronic format and how, if at all, it differs from established copyright law'. A considerable part of Chapter 4 is devoted to this vexing subject.

Chapter 5 then builds on these issues to discuss those inspirational finishing touches by providing practical examples/ideas of things to do with your web pages. We also explore ways that your WBI can be implemented into your overall teaching strategies. The topic of motivation is also discussed, in terms of your own motivation and that of the students.

Chapters 6, 7 and 8 discuss in detail various technical aspects in order that you can physically set up your web pages. In an ideal world, universities would be proactive in providing technical support for staff to set up web pages (Maddux *et al.* 1999: 43). The fact that most are not doing so means that discussion is provided here of software issues, web graphic formats, design considerations and hardware considerations such as scanning and digital cameras. Understanding and using HTML, the standard language of the world wide web, is particularly important and there is a chapter devoted to it and creating pages (see also Scigliano *et al.* 1996). Finally, in this group of chapters, Chapter 8 discusses ways of adding interactivity to your web pages through features such as forms (also quizzes — on which see Kerven *et al.* 1998), easy JavaScript and both synchronous and asynchronous web conferencing. Again, some aspects of these considerations find expression in individual short articles (see Reed and Afieh 1998; Cabell *et al.* 1997 on Java) but here we seek to integrate them into a more satisfying whole.

With your web pages carefully constructed and decorated you will be in a position to test them with students and adapt them accordingly. Chapter 9 deals with such issues, as well as implementation, evaluation and assessing learning outcomes. An advantage of web pages over other forms of learning resource is their capacity to evolve over time and with experience. Not only can a conservatory be added but, unlike with a house, whole rooms can be moved!

This book comes with a glossary of key terms and if you are interested in pedagogical rationales or technical specifications a full bibliography provides routes for further investigations of an area that will become increasingly important.

This book is not intended to be the last word in web design. Accordingly, not all possible angles are covered. Nonetheless, the experience of the authors suggests that careful planning is required. Knocking together a few relevant web sites with no clear idea of how these fit together or how they enhance the student's learning experience is not only unsatisfying but very probably a waste of time.

We hope that readers will be able to use this book to tap into our experiences. We come from different backgrounds and have produced a range of products, based on different academic disciplines and aimed at students at different levels of their university education. We hope that this shows that the material that follows will give hope and inspiration to teachers from all sectors and from all disciplines, regardless of the level of their technical expertise. With time, practice, patience, dedication and perseverance you too can make your own fulfilling, exciting and much appreciated WBI.

chapter one

why use the web for learning and teaching?

This is not the age of pamphleteers.
It is the age of the engineers.
The spark-gap is mightier than the pen.

Lancelot Hogben *Science for the Citizen* (1938)

Many people are quite unsure about how they feel about the integration of technology into pedagogies largely unchanged for centuries in university and other teaching arenas, fearing, for instance, that 'when using these environments pupils may not be well prepared in the wider skills that other...more traditional, teaching techniques and practices can provide' (Coleman 1999). In the case of open and distance learning (ODL) technologies this is true of both teachers and students. McConigle and Eggers (1998) suggest that instructors may feel (consecutively) excited, apprehensive, questioning, determined, over-stimulated, questioning again and exhausted in attempting to integrate newer technologies into their teaching. For students, the 'stages' are described as confused, shocked, timid, frustrated and 'Eureka'; it is worrying that students might arrive at such a climax as teachers find themselves simply exhausted! To summarize, it might be said that for all who encounter ODL the combination is sometimes one of both fear *and* excitement (Williams 1999). Whatever the fears, traditional or accepted methods should not be held sacrosanct; we are talking about pedagogies which some, though perhaps rather harshly, might intimate involve a process whereby knowledge is passed from the notes of the lecturer to the notes of the student without passing through the minds of either. Less harshly, or at least more descriptively, we might say that the traditional lecturer tries to dominate and gain the full attention of the students (and can be legitimately irritated if this is not offered). The lecturer is, or tries to be, in total control of the learning experience and the students are obliged to be passive recipients of information and ideas (and may be legitimately bored if this does not interest them!). Even without deifying technology, there is every reason to suppose that such a situation can be improved upon and, though it can be addressed *within* traditional methods, this is the real possibility for ODL on the web and for excitement rather than fear.

A small caveat should be entered at this point in the sense that simply comparing traditional methods versus WBI implies some claims which are difficult to sustain, and the differences may be easily exaggerated. Obviously, so-called 'traditional' teaching can be made highly student-centred and may encourage very different ways of learning. Similarly, WBI may be unreflective and without time and effort cannot be assumed automatically to provide an advance in teaching methodology and improve student learning experiences (Firdyiwek 1999). Indeed very often, 'instructional designers and curriculum developers have become enamoured of the latest technologies without dealing with the underlying issues of learner characteristics and needs' (Sherry 1996: 337) so that the net effect is negative. This said, there are potential positive aspects which mean that the possibilities of ODL on the web need to be fully explored and utilized. As Relan and Gillani (1997: 43) have put it: 'Having permanent access to a multitude of learning resources regardless of one's geographical location allows continuity in learning and encourages uninterrupted reflection about a topic and revision of one's thesis.'

It is useful, as Boling and Frick (1997) suggest, to pose a number of questions at the outset:

- What problem are you trying to solve? Will another method solve it more easily?
- How will putting parts of your courses on the web benefit your students? What does the use of this type of technology allow your students to do that could not be done before in some other way?

One of the advantages of technology is that effective use of web-based ODL packages can (though a 'combination' model may be used) mean the lecturer relinquishing direct control of the learning/teaching environment. Instead, the onus is placed on students who must interact meaningfully with the materials supplied. The lecturer's role becomes less prescriptive and more that of a facilitator, while the student's role changes from being a passive listener to an engaged learner (Roberts 1999). The world for which students are being prepared requires that they take responsibility for their learning; flexible working practices mean students need to learn in new ways. 'Letting go' encourages feelings of fear and excitement and ultimately reward in more than simply this context of pedagogies but is usually beneficial.

Though this potential may be very real, the fact is that attitudinal studies indicate very varied responses amongst academics of most disciplines (James 1998; Yong and Wang 1996). Some zealously see it as the *only* way to continue to provide a high quality education in the face of increasing student numbers and demands on time, though others are clearly reluctant to go down this road (Rowntree 1998a) and abandon methods which are said to have served well for centuries. The current situation is somewhat analogous to a huge marathon race, like London or New York—some are racing ahead, many are gamely trying to keep up and the majority are a host of struggling stragglers, as well as quite a few spectators too!

However, what is certain is that ignoring technology will not make it go away; even for confirmed technophobes, as one of the authors of this text claims (or at least initially *claimed*) to be. 'Know thine enemy' is a sound axiom in this case and is probably the safest course to follow. We contend that with the aid of books like this, the process need not be overly complicated and that excitement (and reward) can win out over fear. Of course fear can also be overcome by the 'ostrich' route, but it is suggested that putting one's head in the sand is likely to end unsatisfactorily, since only the most conservative would deny that information technologies will come to have a major impact on our lives, including on teaching and learning in the future (Zepke 1998).

The internet is a rich information, communication and research resource for all those involved in education and training (Bannan and Milheim 1997; Crossman 1997; Eklund 1999; Patel and Hobbs 1998; Reeves and Reeves 1997; Stein 1999; Watson and Rossestt 1999) and it *can* be used productively and wisely and even in political terms, progressively, leading to democratic communities of learners. However, care should be taken since the web is medium not magic, technology not thaumaturgy, engine not elixir. It is easy to get carried away — let the web be slave, not master. Rather than assuming that such technologies are the latest expression of social progress, it is possible that cultural forms of knowledge will be lost through the educational use of computers, and this prospect should at least be regarded with caution (Bowers 1998). The work of some suggests that a conservative emphasis and excess faith in science may have implications which are not culturally sensitive (Clark and Estes 1999). Clearly the socially embedded nature of technology is a relevant factor beyond this context, and needs to be borne in mind. On the other hand, neither should one be unnecessarily cautious when technologies are advancing considerably, allowing inventive and creative pedagogies. The need for flexible learning and teaching is increasing (Peters 1998; Williams 1999) and, given a degree of reflection, can overcome certain 'imperialistic' problems. The combination of an information-rich web and your own reflection on learning materials can be the

> 'The World Wide Web provides significant benefits when applied to the classroom. At the same time, instructors, trainers, developers, and researchers need to recognize that the World Wide Web has limitations. Everything that is feasible is not necessarily useful, and everything that is useful is not necessarily feasible.'
>
> BUTLER 1997: 423

web-based packages which this book aims to help you construct and integrate into your teaching.

Assessing your needs

The first thing to consider here is the type of materials you will be using. For instance, are you attempting to make sense of other people's material or are you looking to integrate your own information more fully into a web-*based* package? In either case, we are assuming that the desired outcome is more than simply providing information, which is perhaps the way in which the web is most frequently used, to date, in teaching, or recommended to students. Reading out, or scribbling down URLs in lectures can prove to be highly unsatisfactory indeed! In thinking about such issues of use, you will need to think not only about the level of technology available to you but also how you intend it to be used, as well as the time constraints involved. Excellent WBI packages can be produced without very high levels of technology, although obviously more sophisticated multimedia packages will have certain requirements. Your pages will not necessarily be better because they have more technology, just as your teaching will not automatically improve because you use technology at all.

One thing which should be stressed at the outset is that whilst this book aims to make the setting up of tailor-made ODL web-based packages possible, it does not claim either that doing so will be easy (which is different from saying that it *will* be complicated), nor that having done so enormous amounts of free time will be created. As other studies have indicated, making the transition from traditional to computer-based learning systems is going to involve considerable investment of time on the part of lecturers, especially those new to the web, in the acquisition of new skills and the preparation of web materials (Cornell 1999). Whilst the web undeniably has great potential to benefit students' learning there are, contrary to the hopes and beliefs of many, few short cuts to the realization of this potential (Ward and Newlands 1998). Even where sufficient technical support allows for the rapid setting up of the actual pages, follow-up time will still be required on an ongoing basis, if learning is to be enhanced (Polyson 1997).

Indeed, though we concentrate our efforts here on the conceptualization and setting up of web-based learning packages, the ongoing role of the tutor is absolutely critical to the success of any such development. DIY or bespoke packages give the committed teacher the chance to think through a range of tasks and learning activities which will have the maximum benefits in terms of what they are trying to teach. One important reason for the value of this is that a particular problem of distance packages has been that students, since they are left to their own devices, are seriously impeded by any difficult or troublesome points they come across in their study. Ensuring the availability of guidance is thus crucial to any ODL project (Lee *et al.* 1997) and this militates against the possibility of simply setting up and then leaving students to fend for themselves.

Such support need not be unduly onerous and can usually be resolved through the use of email, or perhaps specially scheduled synchronous web-conferencing sessions in which tutors play an active role, or through asynchronous topic-led conferences. A particular advantage of such sessions is that often it is easier for the lecturer not to dominate these sessions than it is in traditional seminar situations where the tutor is the focus of attention. As one study has put it, 'new technologies now allow for a powerful combination of highly interactive stand-alone material with two-

Untangled web: developing teaching on the internet

way asynchronous communication between teachers and students' (Bates 1997: 93). Ironically, such methods of teaching, though they can take place at a distance, can also bring peers and teachers closer, in a non-literal sense. By breaking down the power dynamics which tend to dominate conventional classroom environments, however well intentioned even the teacher of higher education might be, electronic forms of communication allow even those students who hardly ever contribute in class to enter into a forthright debate with their tutor/s (Harris 1998).

There is some resistance to the very idea of distance learning as somehow not as good as traditional methods, though perhaps those offering such resistance are unlikely to be reading these words (Richards 1997). Certainly such resource-based teaching, if misused, does not provide a high quality education (Bostock 1997). One of the chief aims of WBI in a course, however, is to encourage more active learning and to make students more responsive to what they learn and how they learn, thereby enhancing other areas of teaching (Madan 1996). Giving the students more responsibility does not equate, or at least should not equate, to abandonment. In this sense of encouraging deeper learning, this book hopes to demonstrate web-based ODL's value in a humanities/social science context as others have tried to do for computing, science and engineering (Sloane 1997).

As suggested, there are significant differences between traditional university course pedagogy and effective pedagogy when using technology, which this book will highlight (Schuttloffel 1998). In this context, Jevons and Northcott (1994) suggest a typology of ways in which distance education materials may be used. Before looking at these, it is suggested that not all of these be used. This may be useful in beginning to think about your own needs.

However much it has not always lived up to its promise, the traditional university is founded upon some sense of the need to exchange ideas and information, rather than simply conveying information. In the pure ODL university something is lost and this is implicit in some of the models suggested. ODL, with the support of the traditional tutoring structure, can and should be flexible and expand learning, with online environments helping to establish more constructive learning patterns, not simply eliminating good aspects of what went before (Harris 1998).

Looking at the six-fold typology offered by Jevons and Northcott (1994), not recommended, for instance, would be 'Substitution with face-to-face contact [where] the…distance education materials are used as a substitute for lectures and their effectiveness is gauged by the reduction in class time that is made possible by doing so.' This is a dream for some researchers but, as the authors tellingly, and quite rightly, add, 'when reduction is taken too far…this is regarded as malpractice'. The purely substitutive role of ODL technologies by efficiency-crazed bureaucrats or by researchers with selfish aims is not to be encouraged, and indeed may be counter-productive in the longer term. Furthermore, even where it is attempted as a bona-fide experiment (see Ward and Newlands 1998) the conclusions are a) that time is not necessarily saved due to the increase in tutorial support that is needed and b) that some students experience serious motivational problems once a framework of regular scheduled sessions is removed.

On the other hand, the use of and generation of good learning outcomes with the second category is something with which the authors, and many others, are familiar. Jevons and Northcott's (1994) second category is 'Substitution with different use of face-to-face contact time [where] students are asked to read education materials in advance of lectures or tutorials, which are then used to treat the matter in a different form, or from a different perspective, or to select aspects for special consideration.' In this case students can be asked to consult online materials in advance and do not have the same opt-out possibilities as they do if you have sought to place specific paper materials in their hands, which they invariably 'didn't get' or 'lost'. Effectively this is using web-based materials as seminar or tutorial preparation.

The third category in the typology is known as 'enhancement'. Here, 'lectures are used to introduce the subject matter and the study of the distance education materials is left for the students to undertake'. This model is good, as this book will argue, but direction is needed for the students and careful consideration must be given to the design of the pages. Clear instructions as part of your web-based package are essential. Students are given a degree of responsibility, choice and opportunity which exceeds that offered by a reading list of materials in heavy demand. But, as with reading lists, WBI products are more useful where they are carefully thought out and where assistance/direction is offered.

Category four is termed 'parallel use' and might be quite difficult to justify to anyone assessing the quality of your teaching; 'Students are offered a choice: they can attend lectures, or study the distance education materials, or do both. The distance education materials form back up provision.' Experience shows that it is much better to use category three 'enhancement' or to phrase this fourth category in a manner more resembling the fifth. Jevons and Northcott's fifth category is a 'safety net'. Here, 'the availability of the distance education materials acts as a safety net, which may be valuable for students of various kinds. They include students who miss individual lectures for a variety of reasons including part-time work; students caught in the "note-taking trap" who concentrate on getting it down at the expense of comprehension; and other borderline students who would otherwise fall by the wayside.' Jevons and Northcott add that 'some student counsellors are convinced that the attrition rate is reduced by the safety net'; in this context, category five is much more amenable to defence in terms of teaching quality reviews than making materials purely substitutive or entirely optional.

Finally, number six is described as 'Combination'. In this category, 'over a semester the lecturer may choose a combination of methods. The availability of the distance education materials makes this flexibility possible.' This is basically the model that the authors have tried to use with their web-based pages, though it is worth remembering that 'it is important, of course, for the lecturer to plan well in advance how the materials are to be used so that students can be made clearly aware of what is expected of them' (Jevons and Northcott 1994: 143). In essence, the category of 'combination' implies a level of activity and reflection that the other categories do not; it is perhaps the ideal to which this book aspires and encapsulates the catch-all notion of 'much is possible but make sure it is well thought through in terms of outcome'.

Having said this, using web-based technologies to supplement teaching and to provide a safety net is entirely possible and laudable and this book does not advise getting hung up on perfection, as we re-emphasize in Chapter 9. The Jevons and Northcott typology provides, in any case, a useful framework for thinking about the use of web-based resources (Roberts 1998). The Ward and Newlands study (1998) suggests that students are happy with electronic course delivery as supplementary or support material, and are not excessively worried about looks but rather functionality. It is worth reiterating their conclusion that 'hopes that use of the Web will enable teaching time to be reduced significantly may be illusory' (p. 183). On the other hand, our own experience suggests that hopes that web-based ODL can make teaching more rewarding and learning more interesting are not so illusory.

Assessing your end users

The demand for flexible delivery in higher education is increasing. Flexibility has moved through the stage of the modularization of many degree programmes and has now reached the stage of flexible delivery in terms of these modules themselves, meaning fewer time and/or geographical constraints on access to learning resources. For staff, as discussed above, changing behaviour and attitudes, developed over many years, will not happen overnight, and, as with modularization, may encounter resistance. Nevertheless, a careful assessment of your needs and current technologies may well reveal that these can be of much help to you.

For students (of all disciplines), on the other hand, these developments are happening with *them* and seem less difficult, if that is the correct word (Williams 1999). Even so, any attempt to use web-based technologies in ODL should be fully cognizant that there are currently huge variations in student attitudes to the use of the internet and multimedia techniques, and courses should be long enough to allow all students to get to grips with potentially new and unfamiliar methods and approaches (Perry *et al.* 1998).

You need to think clearly about how you intend your users to learn. What styles will they be expected to employ? What are their levels of IT literacy and motivation and how will learning outcomes be measured? Whilst certain methods of teaching are accepted as normal, there are still, as we have noted above, an enormous range of different attitudes towards WBI, and this includes students as well as staff (Usip and Bee 1998).

The learning that students undertake through interaction with ODL web pages can be of a highly empowering sort. Students often find using the web in their learning exciting and gain increased confidence in the use of computer technology. Web-based projects can be used to encourage active learning by students and 'to achieve the empowerment to acquire knowledge by doing' (Greene 1999). Multiple paths to information and the possibility of non-linear text organization are just two of the important possibilities for students using the web (O'Carroll 1997). There is certainly much evidence to suggest that peer–peer and student–tutor interactivity can be vastly increased by use of ODL web-based technologies (Mirza 1998). However, not all educators are entirely uncritical of educational technology generally (Clark and Estes 1999; Kearsley 1998) and the internet in particular (Kaufman 1998; Siegel and Kirkley 1997).

As mentioned above, electronic communications not only allow more considered responses on the part of students than would be the case in traditional seminars, but also encourage even the most timid of students to express themselves in a more argument-oriented way. This point is highlighted by the transcript reproduced below.

(The debate concerned the extent to which there can be considered to be an environmental crisis, whether it had 'always' existed, whether it was over-stated etc.)

Start: 2.30 pm

14.29:38 Student 1: The economically driven 80s possibly did not want to recognise the issues, but now it seems more socially acceptable to be green.

14.29:41 Student 2: Couldn't it be argued that the problems have been known about for some time but that it is only now that they threaten the 'north' that they have been prioritised?

14.34:33 Student 3: There is a crisis but I think that it is being interpreted in the wrong way in IR. There are more pressing security issues than hippies and tigers.

14:37:09 Student 4: I think that for many, the issue of the environment is now seen as something of a fashion. People just jump on the bandwagon with no real knowledge. This probably has nothing to do with what anyone else is talking about, I'm just having difficulty working things out, so at least I made the effort.

This then led into an hour or so of debate in which half the class joined in at some point. There are several points which can/should be made about this transcript. First, it is not highlighted here because of the great originality of its contributions since this was the first debate of the course. Second, the tutor's own comments have been removed from the above to highlight the types of student contribution. Third, any 'whispered' conversations do not appear on the transcript. Fourth, synchronous web-conferencing is an inexact science. Protocols for use need to be established to make the most of them. Malikowski (1997) provides a comprehensive overview of conferencing issues. In this case these were developed by negotiation with students, with modifications being added after each session. Points three and four are expanded upon later in the book. However, a number of more substantive points need to be made.

Despite the official start time of 2.30 it can be seen from the transcript that certain students are keen to make their initial points early. Student 3 is a frequent 'adversary' of students 1 and 2 in seminar situations and comes in to widen the debate beyond its initial scope to include questions of the purpose of IR. This is done in a reasonably non-academic way, which in this case is fine, but again raises issues of protocols common to forms of electronic communication. Student 4 is interesting. The comments made by this student clearly fit the tone of the debate and, despite the student's own comments/reservations, contribute something to it. Moreover, this is a student who rarely or never speaks in seminar classes and clearly gets a degree of satisfaction from being in a situation where he can at least make the effort, to use his own term. This student participated actively thereafter, including fairly blunt observations about not understanding the comments of the tutor; unlike in a seminar where nothing may have been said, these comments could then be reformulated and explained. Of course the aim is not simply to allow someone to express themselves in this way; by engaging in this debate where more students than normal participate, the tutor is able to draw on comments to entice a broader range of students into the more traditional classroom discussions.

However, there is a need for caution. A study to compare

traditional instruction versus the use of distance learning technologies used an either/or methodology and concluded that the former was more appreciated by students and led to better learning outcomes. In this case it seems as if ODL was used in a purely substitutive fashion. However, this brings us to the secondary conclusion that such technologies are most appropriate when they augment traditional teaching methods rather than replacing them (Thyer *et al.* 1997). Just as the computer does not make errors, but rather human beings do, so web-based packages are not of themselves responsible for ineffective learning but the teachers who have designed them and who support them. Above all, WBI must encourage learners to do something with the information they see. Finally, despite all the positive developments and great potential of ODL and web-based learning, students' concerns about electronically delivered course materials should always be a big consideration, especially if they are involved in assessment (Harris *et al.* 1998). Basically, though, you can probably assume that today's student is more comfortable with technology than those of just a decade ago. You must, nonetheless, be aware that not all students will have embraced technology and some provision must be made to support different levels of student ability. If you are planning to use your web pages as part of your assessment package, it is particularly important that you do not leave any students disadvantaged by their level of technical ability.

Planning for integration

The web pages and their use need to be incorporated into course documentation and dedicated sessions need to be organized to explain them clearly. The integration of instructional principles and strategies with web-based tools is particularly important

(Collis *et al.* 1997). Matching technologies to student need and levels of expertise is recommended (Hodes 1998), though web-based packages can be an excellent way to break technophobic resistance amongst some groups of students, since the technology is not overly complicated and can be related to genuine learning outcomes. In this context, relating the pages to the broader course through directed learning and mock examination questions is also a good idea. Furthermore, the use of technology should not involve too short a time frame. Online environments can inculcate constructive learning patterns if they are introduced clearly and early (Harris 1998) and if they continue to the point where all students are comfortable and where accelerated learning can therefore take place (Leahy 1999).

To a certain extent, it may be best to integrate a less ambitious project to start with and then to let this grow organically. The advantage of a small project initially is that teachers will need to learn how best to utilize technology and how to respond to student needs as they go. The advantage of organic growth is that the many possibilities—interaction with experts and across continents, virtual visits and so on—can be added on as and when relevant (Weinstein and Quesada 1997), or when the instructor/designer has time! The tendency to charge in enthusiastically is one that should be guarded against. On the other hand, where institutional support is sufficient, the internet can be 'embraced as a medium for supporting students, tutors, academics and administrators throughout the education process' and via an holistic approach to integrating technology into the teaching process (Thomas *et al.* 1998).

Perhaps the key point about integration is to introduce a little conservatism, along with ideas of progressive teaching methods, in an era of radical change. If you are working more or less alone it is better to plan a little properly, than to half-finish an ambitious project which is full of problems. Especially if you need more technical knowledge yourself, there is little to be gained from what might be termed running before you can walk.

Evolving and quantifying learning outcomes

These need to be effectively monitored. Web pages have enormous potential in terms of acquiring a range of key skills; pages can be important sources of up-to-date information via individual research. They can also give rise, however, to a whole range of reflective and group work tasks. It is important to state clearly to your students, both verbally, if possible, and in course documentation, why they are using the pages and what they are learning. It is also useful for this to be reiterated within the pages themselves so that students are explicitly aware of how your web pages are contributing to their learning experience. Since there is already widespread awareness of the need to teach key transferable skills and wide variations between the types of skills that different disciplines are able to teach, this aspect is not covered here. Needless to say, apart from computer literacy skills, ODL web pages have the potential to contribute in teaching a wide range of skills, from research and planning through to communication and group work.

The learning outcomes of a WBI package will be based on the same criteria as those used in identifying and measuring learning outcomes in other, more traditional, delivery systems. Much will depend on the level of the target audience, how much the WBI is integrated into the course, module or whatever. There will be a need constantly to monitor the situation. Not only will you want to measure students' progress but also your own as you acquire new skills and build up expertise. There is also a need to attempt some evaluation of any added benefit derived from using the web, as opposed to other delivery mechanisms. At the outset you must ask yourself—why use the web to do this? There might be personal, economic, political, administrative or technical reasons why you or a team might wish to pursue WBI. But are there pedagogical reasons? It is wise to heed the words of Luchini (1998: 13) when she warns that while the web has great potential, 'it is obvious that caution must be exercised when integrating technology and pedagogy'.

Conclusions

In terms of technology and teaching, the optimists are too optimistic and the pessimists too pessimistic. What this chapter should have served to emphasize is that developing web pages is something that will require careful planning and effort and that the development of WBI products will not, fortunately, reduce the need for teachers nor, unfortunately, reduce workloads, especially since electronic course delivery relies on quick response times by the tutor and is time-intensive. Although others have drawn different conclusions in similar situations, we believe that technology-supported ODL needs a level of support which is not simply going to free up research time, though ODL may, of itself, contribute to research aims (Roberts 1999). If managers or teachers were to see this as a solution to long-term financial or work pressures they would be seriously mistaken. It is not just that so much effort is expended in the setting up of pages, though it is. It is not simply that updating of pages is as important and time-consuming as updating other types of teaching materials, though it is. Neither is it simply that at an institutional level a great deal of financial commitment may be required. It is more than this in the sense that active engagement with the students is still needed; the students must not feel abandoned and isolated in a virtual world, but linked to tasks specially prepared for them and able to 'touch base' as required. If this is done then the web can help to facilitate new types of learning creating non-linear learning patterns, giving the students time to pace their learning and thus providing the time/space for the students to comprehend, reflect, draft, analyze and check in ways that have not previously been possible.

'As the educational use of Web-based technologies becomes widespread, distinctions between distance education and classroom-based education may become less apparent.'

HILL 1997: 79

The problems, then, of ODL projects on the web are largely of time and of successful integration. To solve these problems requires not only a great deal of thought and planning but also other resources. Thus the potentials of ODL on the web, though enormous, are largely pedagogical rather than time or money saving, as we will see.

chapter two
planning the information architecture

Daedalus, an architect famous for his skill, constructed the maze, confusing the usual marks of direction, and leading the eye of the beholder astray by devious paths winding in different directions. …Daedalus…was himself scarcely able to find his way back to the entrance, so confusing was the maze.

Ovid *Metamorphoses* (Innes 1955: 183)

To create a web site, all you need is some material, some HTML-scripted pages and very probably several images of some description. To publish your site, you need access to a web server, whether within an educational institution or through an Internet Service Provider (ISP). But beyond these elements, there are numerous paths to a successful learning-oriented site. Sound planning will help you get there with a degree of composure.

Preliminary planning considerations

Whether your envisaged WBI will comprise a page or two, or hundreds of pages, pre-planning of the architecture, of the content and of the facilities is essential. It saves time by identifying problems before they arise, will aid you in resource planning and, importantly, greatly improve the likelihood of creating a coherent site that is intelligible to the students.

There are a number of factors to be considered in planning a site, irrespective of the subject area it will explore. These can perhaps be best approached as a series of questions that will help you develop a clearer image of the parameters within which you will be working.

- **Consider the subject to be covered**
How best can you exploit the medium of delivery? Will you be using case studies? Links to other sites? Online discussion? Multimedia? Later chapters of this book will give you some ideas, or perhaps you have already seen some interesting experiments on the web that you think would work well within your subject area.

Some very useful pages have been created from simple technologies using good content and a bit of ingenuity.

- ## Consider what **exists already**
Is the very thing you need online already? Are there substantial resources on the web which you can utilize? There is no use reinventing the wheel, after all.

- ## Consider **who will be using** the end product
At what stage are they in their studies? How much do you know of their learning styles and habits? How developed are their IT skills? 'First contact' with the material must be a positive experience, or the students will be reticent to use it again. If it is likely that they will have mixed IT abilities, you must develop a strategy so that those with fewer IT skills will have a chance to experiment in a semi-supported environment. Although feedback from students has indicated that they prefer to be left alone with the computer-delivered material (Keady 1999), they may well need an initial session where help is at hand. A model which often works well is that of one timetabled introductory class in a computing room, where the students begin to use the material, but their tutor is present in the room should they need help. Coupled to this, we have found that an 'orientation' section built into the web site and the opportunity to use 'drop in' IT help sessions both provide some support for longer-term queries and for more IT-savvy students who might not attend that first session.

- ## Consider how the students will **access** the material
When and where will they gain access to the material? To what technology do they have access? The technology that the students will use strongly influences the developer's choices on what technologies can be employed to deliver the material. For example, there would be no point in including photographs in the web site

if most or all of the students would be using a text-only browser to view it. But if only one or two per cent were, you would probably include pictures, but make sure there were good text descriptions, too.

Computers and software provided in Universities may well be below the given 'standard' specification of the time, or in well-funded Faculties, they may be considerably better. And although the great advantage of the web is that it can deliver to all platforms, in reality there are differences in how different computers and browsers will display a given web page. Time spent investigating the facilities available to your students is time well spent and may save serious heart-searching and backtracking later. A project that relies on video or audio clips of interviews with subjects will not be successful if the computers to which most students have access are in labs with no sound provision.

Increasing numbers of students have internet access at home, and will experience the material downloaded slowly through a modem. It is important to consider the students who use the site in this way. Although they may well have access to superior computers, they will be connecting via modem and will very probably be paying for the telephone calls. These students will not wish to spend a great deal of time online, moving around the site. A more common pattern of use for this group will be to log on to the site, search quickly for the information they currently need, save or download it, then log off.

In May 1999, visits recorded to a Faculty's server demonstrated that more people were using a previous generation of browser than the very latest version. Many of these visitors would have been students using the University's computer rooms.

You do not always have to design for the lowest common denominator, the lowest specification, the most basic browser or machine. But it is a good idea to design for what most of your students will be using, if a clear pattern emerges. These 'most likely conditions' will provide a baseline for what you can deliver, although it is reasonable to expect to have to make some alternatives for students with specifications below this level. It is a question of balance. Table 2.1 demonstrates browser 'obsolescence' on one server.

Table 2.1 Web browsers used by visitors, May 1999

Percentage of visitors	Browser
30	Explorer 4.0
20	Netscape 4.0
15	Netscape 3.0
12	Other browsers
11	Netscape 4.5
8	Explorer 5.0
1	Later and earlier versions of Netscape and Explorer

- **Consider how the end product will relate to the module or course**

Will it be integral? A course companion? Entire of itself? An add-on? For distance or for local learning? How you are planning to use the web site will, of course, have significant influence on the complexity of the content, on what utilities will be needed (and therefore, what technologies) and on how much built-in support the students will need. You will also need to consider the practicalities of the students' use of the site—for example you may decide that you need to make room in the timetable, or accommodate some changes in existing teaching or assessment techniques. It is our experience that students need clear incentives if they are to embrace learning on the web—creating a site, inviting the students to use it as part of their studies, then carrying

on without appreciable further reference to it is unlikely to
encourage them to make any initial effort.

- ### Consider the **learning outcomes**

What specifically do you want the students to learn from the
material? What information, what skills? How can the learning
outcomes be measured? How will you encourage the students to
engage with the material? Learning outcomes should be clear to
the students and explicitly stated. As with other learning
techniques, if the students can see a clear benefit or added value in
using the site, they are likely to enjoy a more positive experience of
the material. It can make the difference between the students *using*
the site and them *engaging* with it. Evaluating learning outcomes is
discussed in detail in Chapter 9.

- ### Consider **assessment**

Will the material be assessed? In what way? Formally or
informally? Self-assessed? As part of the site, or as part of
conventional assessment? You will probably wish to assess the
success of the learning outcomes associated with your site and you
will certainly wish to gauge in some way how the site performed.
You may also wish to use assessment as an incentive to the
students to work through the material. How formal the assessment
you wish to use and how much you wish to utilize technology in
the process will have ramifications for the project and its
integration. For example, you may wish to adapt the assessment of
an existing module or course to include the web-based material —
promise a related examination question, perhaps, or an essay
topic. Or you may wish to use the site as a revision aid for class
tests. If you would like to incorporate assessment into the web site
itself, you need to consider the resources available to you. A self-
assessed or an auto-marked quiz is very straightforward to
implement. These are useful for revision and reinforcement and

the students appear to enjoy them very much. You can also use various methods for submission of informally assessed work, but if you wish the students to submit formal assessments electronically, you must give some serious thought to security.

The assessment of web-based learning is discussed in more detail in Chapters 5 and 9 and simple techniques for developing web-based assessment are described in Chapter 8.

- ## Consider **future development**

Will the material be expanded? How will you obtain useful feedback? You may choose to allow your site to grow organically — which may have benefits for certain types of subject matter — but if your site grows very large, or if you need to present material in a structured way, it is advisable to plan for any future growth. This approach will often save you from having to restructure your site significantly, later on. Although the much-reviled 'under construction' pages are best avoided, you will nevertheless need to consider building 'invisible' expansion space into your site.

Feedback will, of course, be useful in assessing the success of the site and in planning any future changes. You may well wish to build some provision for feedback into your site, or you may decide to obtain feedback in paper format, perhaps as part of wider feedback on the module or course. If you decide on the former, you will need to decide what form this will take — a popular format for feedback requests is an electronic form that is emailed back to you, or to a database. (Forms are described in Chapters 7 and 8.) Alternatively, you may decide to post some questions on a web page or message board and invite replies by email, although this will probably result in a lower response rate, as it is slightly more effort for the respondent.

- **Consider the resources to which you have access**

Do you have the technology to produce what you want to do? Can you get any help? Do you have the time? Contemplation of the technology to build web sites is often initially the most off-putting factor to lecturers who may be unfamiliar with the developmental terrain. In reality, this is often not as limiting a factor as people expect. Much can be achieved with the simple technologies described in this book.

You will also need to investigate how your site will be delivered — as we are considering web delivery here, you will need to find out whether you have access to a web server in your institution (and who administers it), or whether you will need to use a commercial or academic Internet Service Provider (ISP). Available resources will vary considerably, so ask what facilities are available to you, how you can upload and update your site and what the server specifications are.

Questions to ask your server administrator or ISP

- Is there a size limit on what and how much I can upload for my site?
- Is there a limit on the daily traffic allowable to my site?
- Can I upload my pages myself? When will changes I make appear?
- What facilities does the server have? A search facility? Form-processing capability? What Common Gateway Interface programs (CGIs) are available? How do I make use of the available facilities?
- Will I be able to run my own CGIs from my own directory? (Probably not, in fact.)
- What platform is the server?

- Is it a fast or slow server? (This question may well offend!) On a fast or slow network or connection?

The answers to these questions will help you plan a more successful project. They will guide your decisions later on. Looking at the example in Table 2.2, some conclusions about the shape and content of the material can be reached straight away. The exercise does not take long—you probably know much of the information already—but it is very worthwhile to take the time to go through it formally. Like all preliminary planning exercises, it helps to crystallize the conditions with which you have to work and will provide a useful reference for the project's duration. Not only does this preliminary planning help in weighing up the cost to benefit relationship, but also keeps development costs down as problems can be foreseen and avoided.

Table 2.2 Planning considerations example—the *Social Geography and Nottingham* site

	Considerations	Implications
Preliminary		
Subject	Case studies of the local area, which will illustrate the central themes of the course.	Data easy to collect.
Medium	Photographs, walk-through panorama. Use linkages to engender a sense of scale. Online self-assessed revision. Ongoing updates of material.	Some of these will give large file sizes, making slower loading pages. Do students have access to the *QuickTime* plug-in to view panorama graphics?
Existing material	Nothing suitable online. No significant resource, but some sites suitable to cite as 'further reading'.	

Students		
Stage	Level 2, from various undergraduate courses.	All have taken Geography modules before. Assume most will have some experience of directed learning.
Access	Some will access the material from home, some from libraries, but most from University computing resource rooms. Most will access the material during the network's peak use hours. Must consider future students who may have specific access needs.	Bandwidth will be a consideration. Large files may take a long time to download. Note: test this at peak hours. Where sound is included, it must be supported by text transcripts. Site must be made friendly for screen-readers.
Technology	Specification for University machines: Windows NT, no sound card, 256 colours, 640×480 screen resolution, 15 inch screens, running Netscape 3. No control over the plug-ins loaded.	Audio material cannot form a major part of the material. Visual design must take screen resolution into account. JavaScript must be limited, or alternatives provided, Shockwave technology cannot be used, and the HTML must be degradable to the older browser (so Style Sheets will not provide an advantage to formatting).
IT skills	Mixed.	Introduce the students to the material in a timetabled session, with staff at hand to help. Include orientation section in web site.
Learning styles	Mixed.	Ensure there is a range of material and activities to suit different approaches.

Learning		
Relation to module	Will run parallel to the second half of the module. The material will reflect the themes discussed in lectures and will feed back into seminars. A question on the material will be included in the end of semester exam. Students will be free to use the material when they please, but will be strongly encouraged to work on it for an hour a week over 5–6 weeks.	Organize material so that it can be worked through in hour-long chunks, if the student wishes. Include sample exam questions for revision.
Learning outcomes	Students will acquire and reinforce skills in using a web browser and familiarity with the internet. They will learn techniques of self-paced learning, with varying degrees of direction. Students will learn to apply theory to particular case studies, gain an appreciation of some of the key issues in the social geography of Nottingham and reach an understanding of the importance of scale in the study of social geography. Learning outcomes will be measured by means of feedback and group contributions to seminars, by completion of the online feedback form and by an exam question.	Outline learning outcomes section in introduction. Include activities through which the students can measure their progress. Online forms required for feedback.

Assessment	Formal assessment will be by exam question. There will also be self-assessed and auto-marked tests and questions built into the online material, so that the students can monitor their own progress.	
Development	New case studies may be added in the future. Depending on feedback, functionality of the material may be developed, including more online assessment and possibly online discussion in workgroups.	When planning the structure of the material, build in provision for planned later expansions.
Resources	Technology and knowledge is sufficient to produce material which rests on web pages, multimedia and forms, but complex programming would need to be outsourced. Project will be served from the Faculty's own web server.	
Time	One hundred and twenty hours can be allocated to the project, over the summer vacation period.	Probably the most influential consideration in what can be produced. May need to scale down and plan to expand later.

Planning the content

Just as the truly wise never attempt to build flat-pack furniture without checking they have all the pieces, so it will save time and pain if you make a full list of all the content you would like to include in your site. John Shiple, writing for the web developer's magazine, *Webmonkey*, identifies two main types of content — *static* and *functional* (1998) — highlighting the need to list both the material content of your site and the practical functions, or machinery needed to deliver it. So, for example, you might decide you want to include quotations on a particular topic to highlight certain themes in the material you are presenting. You would like the students to be able to elect to view these while they are reading the 'main' content of a page, and you would like the quotations to appear in a little 'popup' window of their own. Your content list for this section might look something like that shown in Table 2.3.

Table 2.3 Listing content

Material content	Functions
Textual introduction to deforestation section. Quotations from X and Y on deforestation	JavaScript to call up, size and close a subsidiary window

If you are planning a large site, it will be a long list, but it will save you time later on and will help you prioritize your time. In the example above, you may decide that given the time that you have, working on your JavaScript 'popup' windows is less of a priority than the electronic feedback form you need to elicit response from the students. At this point, you also need to ascertain if the function you want to include is possible, given your resources — later chapters of this book will help you here. Likewise, listing out all the material content will help you decide whether or not you have time to collect it all.

Mapping the structure

'Cognitive psychologists have known for decades that most people can only hold about four to seven discrete chunks of information in short-term memory. The goal of most organizational schemes is to keep the number of local variables the reader must keep in short-term memory to a minimum, using combination of graphic design and layout conventions along with editorial division of information into discrete units. The way people seek out and use information also suggests that smaller, discrete units of information are more functional and easier to navigate through than long, undifferentiated units.'

LYNCH AND HORTON 1997

Having compiled a thorough list of all the content you intend to include, the next step is to organize it all into sections and decide what will connect to what. Group together material that logically works together. Your rationale might be data or activity based; it might reflect themes or levels of understanding. This grouping of information will form the basis of your whole site architecture. How you *structure* your material is crucially important in encouraging the students to make meaningful interactions with it.

We have experimented with various site structures, offering different pathways or arcs through the material (Table 2.4). These have included straightforward linear paths, open forms and modular structures. The structure you choose depends on your material, your students and the overall aims of your site. Observations of students using distance learning material indicates, perhaps not surprisingly, that for a large site a combination of these approaches is appropriate.

Linear paths

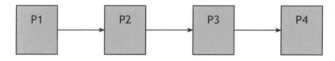

This is the novel of site architectures. The student's progression through the linear path is heavily controlled by the tutor. The route can be a straightforward page to page line, or more complex, perhaps allowing the student to skip sections. Linear structures work well for small sites that have a focused and specific learning objective in mind. They can be useful as a 'way in' for students who are new to the web environment and for students who initially lack confidence in their own learning decisions. The fixed structure not only acts as a guide, but when followed and completed, offers a sense of achievement and satisfaction. Linear paths are, therefore, more suited to directed rather than autonomous learning. Their disadvantages are that they may be too constraining for experienced and confident students; if over-used, they may not encourage autonomy and they can become unwieldy if used for large sites or long narratives.

Open structures

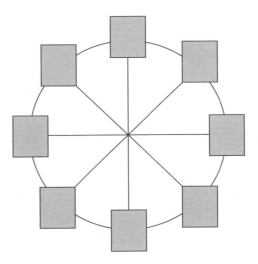

Free or open structures are the web equivalent of a pick 'n' mix confectionery stand. They are unrestrictive and offer many choices of approach to a site's material. Many other points, perhaps even the whole site, may be reached from any given page and no clear route through the material is presented. The open structure allows students to follow their own interests entirely and to take responsibility for their own learning choices. It is more suited to experienced and confident students, who have skills in using the web and in autonomous learning. This structure has been utilized for creative writing and fine art sites, where an unrestricted and individual response to the material is sought, and this is their main strength. The drawback of completely open sites can be that without any guidance, students may become disoriented and discouraged. If the success of the learning outcomes depends on students engaging with all the material in your site, an open structure may not be in their best interests.

Modular structures

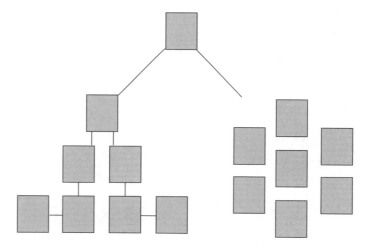

The modular approach falls somewhere in between linear and open. Modular structures offer the choice lacking in linear paths, but also add stability which can be absent in open structures. The student is offered a choice of activities or material, but then works through individual sections in a more linear pattern. You can opt to offer suggestions and advice as to suitable paths, or flag different groups of material as core or supplemental. The different areas can be made to the same pattern, so the students know what to expect and can plan their learning. This structure is useful for a class of mixed levels of experience of the web and distance learning and so has the widest application. A drawback of modular structures is that an unvaried diet of identically structured material can both dull the palate and discourage curiosity.

Narrative arcs

If you decide that your web site needs a more open or a modular structure, but want to offer some informal guidance, a powerful technique is to use narrative arcs. These might comprise themes or motifs that recur throughout different areas of your material, setting up resonances between different sections and encouraging the student to make connections. Narrative arcs are less obtrusive than straightforward instructions at the outset, but can nevertheless provide structure and coherence to your material. The drawback of using this technique is that you can unduly influence the inferences the students make from the material. You may decide this danger is outweighed by the benefit of offering signposts to those students with less experience of autonomous learning.

Thought paths

Another approach is that of using hypertext links to offer 'thought paths' through your material (Whalley 1995: 26). This is one of the major advantages of using the web as a means of delivery: 'It is possible to provide a very rich learning environment using the Web. Hypertext links provide an advantage that a conventional course (or course notes) cannot easily provide. A *thought path* which helps to reinforce linkages can be suggested which is not easy to do in conventional book form (or lecture notes for that

Table 2.4 Structures in practice

Site	Description	Structure
Learning to use the web	Small site, for use during one two-hour class and for solo revision later. Aims to give level two history students the skills to use the web effectively in their research. Students are expected to have different levels of knowledge with regard to using the web.	Essentially **linear**, but with three different jump-in points, depending on the student's current knowledge.
Environment and IR	Distance learning companion, running alongside a seminar and lecture programme. Aims to encourage wider exploration of selected themes from the course, using web resources. Part 'gateway', part launch-pad for discussions, some of which take place online.	Fairly **open** structure, modified by short introductory sections, site map, frequent popup menus and clear signposting. Associated conference board and discussion room.
Social geography	Distance learning element of traditional module, to be worked on by students in their own time in weekly chunks. Aims to illustrate themes of the course with case studies. Ties in closely with the lecture and seminar programme and with the module assessment.	**Modular** structure, with each section offering a choice of linear or open paths through the material. Detailed site index allows students to return to a page, even though it may be in the middle of a narrative.

matter).' When putting this into practice, be aware that planning thought paths through your material can lend a more open structure to your site than you had intended. Many complex sites employ links on every page to suggest pages of related interest, and these are most successful where the links are included as a supplement in a side bar, rather then embedded in the actual textual content of the page. Making links directly from a word in the middle of a sentence can be both liberating and frustrating for the reader.

Learning styles

Although subject to doubts in fields within psychology and education (Reynolds 1997), the concept of 'learning styles' has provided many educators with a basis from which to develop their courses. If you subscribe to the usefulness of learning styles, then you can quite clearly utilize their framework in planning the structure of your online material. In fact, one of the great advantages of the web is that material and activities can be shaped and linked in different ways to serve the needs of different types of learners. Stefanov *et al.* (1998) describe the successes of an internet-based distance learning business course which, through its design, allowed learners to take responsibility for their own learning and follow their individual cognitive styles.

Successive adaptations of Kolb and Fry's (1975) experiential learning styles have delineated several types of learning. Many writers have commented on the applicability of these styles and it has been noted that, rather than there being different discrete types of learner, an individual is likely to adapt his or her learning style to fit a given environment (Tennant 1997). Feedback that we have obtained from students using web materials has indicated that the majority of them show a marked preference for active learning styles in this environment.

Practical considerations

A more prosaic influence on structure is that many students will very probably print out sections of the site to read off-screen, particularly if you plan to include a significant amount of textual content. You might want to consider building in 'print' versions of parts of the site, perhaps as downloadable documents. You also need to ensure that the students do not become lost in the site, whatever structure you opt for. (Navigation, menu depth and other 'ease of use' design considerations are discussed in Chapter 3.) You may decide that you need a whole, parallel, text-only version of your site, although if this can be avoided by careful design, it will save you a lot of work.

Storyboarding

Once your ideas have been sketched out, you may find that formalizing or mapping the structure in some way will be of benefit later—not only in those moments of desperation when you wonder why you began, but as a general lodestone. It will be particularly useful if you are building the site with a team. A map of the architecture of the site could be a flow chart, a storyboard, a tree, or whatever feels appropriate (Figure 2.1). However you can best envisage your site, a pictorial representation will help in envisaging the relationships between its various elements. It is also a good idea to show your completed map to someone else, particularly if you are working on your own. They may make entirely different associations from the ones you have mapped out. Boling and Frick (1997) recommend building a card index prototype of, particularly, large web sites at about this stage and using it for early user testing. This is an interesting approach to consider, as they found that testers were much more likely to be forthcoming about areas of difficulty than they were in a beautifully laid out, 'finished' site.

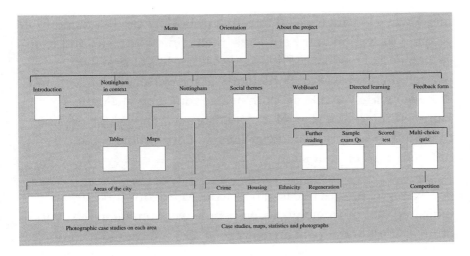

figure 2.1 an overview of the planned site

Patterns of use

The ways in which your students will use your site are difficult to predict, but testing, feedback and judicious use of other people's experiences can assist. For example, when we planned one distance learning site, it was envisaged that students would work online for about an hour a week, in one session. In reality, the students logged on for much briefer sessions (of around 15 minutes), printed some pages out, then returned later to work on the exercises associated with the material. So it would be sensible to redesign the site with this use in mind. This pattern of use also has implications for resource planning, if you intend to build a site for local learning—for example, would there be any point in booking a computer room for a solid hour a week? Or perhaps the very act of making a timetabled room available would influence the way the students used the site?

The pattern of use just described was determined from a combination of server-generated statistics for the site and formal feedback returned from the students. These methods are two of many ways to gauge use. Interesting indicators can be gained from visual observation of the students interacting with a site and from casual, verbal feedback. Published case studies of previous projects are very useful, particularly at the outset of a project, and these are widely available, both in electronic and analogue journals. You might also refer to statistics with a wider base, like those produced for commercial sites (but bear in mind the different aims of your site).

For example, figures released by Media Matrix for June 1999 indicated that during days when people went online in the course of their work, they spent an average of 43.3 minutes logged on (1999). However, people only went online at all in their workplace, on average, on 13.1 days of the month. The figures were very similar for home use. This suggests that the time people are prepared to spend online is similar for work or for leisure—perhaps people have a web-saturation point, whatever the type of use.

Further reading
Case studies in:
Khan, B.H. (ed.) (1997) *Web-Based Instruction*, Educational Technology Publications, Englewood Cliffs, NJ

Schroeder, W. (1999) Steering users isn't easy, *View Source Magazine*,
http://developer.netscape.com/viewsource/schroeder_ui/schroeder_ui.html

chapter three
designing a successful interface

Suddenly she came upon a little three-legged table, all made of solid glass; there was nothing on it except a tiny golden key, and Alice's first thought was that it might belong to one of the doors of the hall; but, alas! either the locks were too large, or the key was too small, but at any rate it would not open any of them.

Lewis Carroll *Alice's Adventures in Wonderland* (1865)

F riends of the authors who make a living as designers often point out that design is a specialized discipline, for which they have had many years' training, and that it is impossible to communicate the intricacies of good design to non-designers in a couple of books or seminars. They are absolutely correct. This being so, it is nevertheless not entirely helpful when you are trying to produce a web site for your students, you have no budget and you would just like a bit of advice. So here we offer some, based on our experiences and with apologies to our friends who work in design.

'The "interface standards" of books in the English-speaking world are well established and widely agreed-upon...Every feature of a book, from the table of contents to the index and footnotes has evolved over the centuries, and readers of early books faced some of the same organizational problems facing the users of hypermedia documents today.'

LYNCH AND HORTON 1997

Many of the major elements of a successful interface between content and user are, of course, integral to the architecture of the site. But a usable interface also encompasses the visual design and methods of navigation. What constitutes 'successful' in both cases is fiercely debated, and is largely a matter of priorities. A useful rule of thumb for most distance learning purposes is to create a style that promotes readability and, while being visually attractive, does not subsume content to decoration. Good design makes a *usable* product.

Navigation

Successful navigation relies on both the architecture of the site and its visual design. The means by which the student is signposted round the material, and the way in which the site, essentially a series of files, becomes a coherent entity, depends on creating a convincing and understandable metaphorical environment.

The phrase 'web page' encourages comparisons with the more

familiar environment of a book or magazine paper page. It is an analogy that can be both of benefit and a disadvantage. For thinking of a web page in terms of a paper page can, has and continues to limit the way in which we build sites. A column of text here, a picture there, number them one to ten and voila!, a document that differs from its paper cousin only in the way it can be delivered, and offers little added value for all your hard work. The converse of this limiting influence is a liberating one—as a familiar metaphor, the 'page' is very useful in aiding the student in making a quick connection with your material, or, if the student is new to the web, in ameliorating the disorienting effect of entering an alien environment.

So not only will a successful site offer an environment that encourages engagement with the content, but also the means by which the student moves around it will be as effortless as possible. Site navigation provides the interface with your material and will **communicate** the structure of the material, encourage **curiosity** and exploration of the site and, importantly, will be **consistent** throughout.

> 'Thought is shaped by tools.'
>
> PAYNE 1991: 128

Good navigation is as intuitive as possible. At any page in your site, the students should know where they are in relation to the rest of the site, they should know where they can go and they should know how to get there. (User testing is very useful here, as what may have become familiar to you while building the site may be less comprehensible to someone approaching the material anew. Testing is discussed in Chapter 9.) The students should also be able to understand the navigational devices you employ, be they text or symbols, without explanation. Familiar symbols in Western society, and therefore popular navigation metaphors, include tape cassette/VCR controls; even represented crudely they have some meaning:

Whatever navigational system you decide upon, do not allow it to become too obscure or self-referential and if it relies on graphics, provide alternative text links somewhere on each page.

Table 3.1 Metaphors that have been used in web sites

Metaphor	Organization signifiers	Navigational devices
Book	Material is organized into chapter-like chunks, signified with running headers and page numbers.	Contents page, index, clickable page numbers.
Desktop software	Files, directories and dialogue boxes.	Tool bars, menus.
Building	Rooms and storeys.	Floorplans, doorways, department store listings.

Organization at site level

Easy navigability and access to the material, of course, also depends on how the site is organized. Your information architectural plan already forms the basis of your whole site organization rationale, and you may have decided upon a suitable metaphor with which to shape the layout of the site. You also need to envisage how the students may move through the site and how their behaviour will influence, and be influenced by, the design of the site. This relationship should, of course, be as positive as possible, and to this end you should consider the actual building bricks of the site. What menus will you need? Will you provide a site map? An orientation area? A search facility?

You will not know exactly how the students will move around the site until you launch it. Although the user-testing discussed in Chapter 9 will help you fine-tune the site, here at the planning and building stage you will have to rely on your own experience of student behaviour and your subject material, and on others' accumulated experience of the interaction between humans and web sites.

First, some thought given to how the students are most likely to move through the site may be helpful in building your site. Which are likely to be the most popular pages? And what probable routes are there to those pages? Is the route too long? Are there any strategies for shortening the routes to the most popular information, but without degrading the architecture of the site?

The most frequented routes, or 'mean paths', through a web site are often calculated by statistics software on the hosting web server, and can be analyzed together with the duration and 'session depth' (the number of pages used) of the average visit. If you can get access to some of these statistics for existing sites with similar aims to your own, they will act as a guide. If not, you can still achieve similar conclusions by thoughtful estimation.

Some accumulated observations, drawn from various sources, including our own experience, may provide illumination here. The behavioural observations in Table 3.2 all influence the way a successful site is organized — some represent the frailer aspects of human nature which must be acknowledged, even if they are unpalatable. Some of the statements in the table have contradictory implications (like those that influence menu depth and page length) and so the most that you can hope to do is decide on your priorities and aim to strike a balance. There are no perfect, ideal solutions, and web designers have been known to argue at length, and with some intensity, about optimum page lengths and menu structures.

Table 3.2 Behaviour and site organization

Behavioural observation	Implications for site organization
Something like 85% of people will not read all the text. Ideally *your* students would read all your information, but in reality they will not.	If there will be a significant amount of text, consider information hierarchies that allow students to read at different depths and home in on areas of interest.
Around 10% of people apparently never scroll down a page at all.	Essential information should be in the top part of each page. Very long pages should be used with caution.
People find that reading long documents on screen is uncomfortable.	Be wary of long documents and consider providing a version for the student to download and print.
People are impatient when waiting for a page to download.	Pages cannot be very lengthy, or if they are, the information therein must all be relevant and so worth waiting for. Subdivide information and use menus, so students can select what information they need to download.
People do not like to traverse too many pages on their way to the information they want.	Avoid over-complex menu structures and subdivision of information into too many short pages.
People like clear signposting, but may choose to ignore it.	Menus at critical junctures will be useful, but should not be essential.

Lynch and Horton (1997) argue for page lengths of not more than one to one-and-a-half screens—this is eminently sensible advice and a reasonable guideline, unless you can think of compelling reasons to do otherwise. A compromise that works for many sites containing long articles is to divide the text between several linked web pages, then to offer a 'full' version of the document for printing. This print version may be a long web page that opens in a new window, or it may be a document to download and open in a document reader or word processor.

Strategies for shortening paths

How can you shorten the paths from related material that resides in different sections of your site, but which the student may want to use consecutively? Or how can you ensure that a student returning time and time again to a site will not have an unbearably long 'click stream' back to the material they were last reading? There are many strategies available:

- Provide an **orientation** page. An introductory area, clearly marked as a good place to begin, can not only offer advice on how to approach the material, but can also provide a guide to the navigation and structure of the site. An orientation page can also offer (or link to) an introduction to the web and to using a browser efficiently. If you highlight here the usefulness of marking pages as 'bookmarks' or 'favourites', the students will be able to mark and return to the page where they were last working. Do not depend on the students' use of bookmarks, however—not just because some students may not read the orientation section, but because the computers they use to access the web may make no provision for making and exporting bookmarks (particularly if they use institutional machines).

- Consider **restructuring**. If your mean paths analysis throws up many linkages between materials that are notionally 'far apart', a review of your architecture might prove useful. If a better way of organizing the site transpires, then rework the architecture. You are not losing a great deal, because you have not physically built the site yet—far better to do it now than later! If, however, you are still convinced that the current structure is the best way to organize your material, then keep it as it is and find another solution.

- One such solution might be '**related interest**' areas, which act as an aside from the material the student is currently working with, but nevertheless signal that related information is available in another area. Examples might be a short linked list to related articles on the site, or a link to past examination questions in this subject, or to activities related to the current topic. Make sure the students have a clearly signposted way back, if they choose to follow one of these links.

- A variant of the 'related links' solution is to utilize **drop-down menus** which contain lists of other pages. These can often be found on the front pages of sites, providing a 'short cut' to popular pages. These drop-down menus (often called 'select and go' menus) have the advantage of taking up little space on the page and their use is intuitive for students with some IT experience. The drawbacks of drop-down menus are that they require use of scripting beyond HTML—like JavaScript—and if a great number of links are listed in the menu, the page will load more slowly.

- Provide a **site map**. This might be a very detailed listing of all the pages in the site, like a directory, a visual 'plan' of the site, or a quick mnemonic in a floating palette or window. Site maps can vary in detail and execution, but their general intent is to provide a sense of direction in the construct that is the web site. A detailed directory of the site is particularly useful for students returning for successive study sessions, as they can quickly jump back to the last page they were using, without threading their way through a potentially lengthy path. For the same reason, site maps are useful for revision—especially if you have included, say, tables of numerical data, which someone might conceivably want to look up quickly.

- Allow the students to perform a text or keyword **search** of the site. Search engines can vary from dedicated software packages with multiple features, through extensions or plug-ins for server software, to CGI scripts that will run simple searches. Simple search scripts can be found on the web to download, and sometimes their authors make them available free. If you do download a free script, make sure it comes from a reputable source and check it over carefully before installing it and running it on a server. (If you know you can run your own CGIs on your web server, it is nevertheless polite to ask your server administrator before you do so.)

Free search scripts at time of writing
at *Matt's Script Archive, Inc.*
http://www.worldwidemart.com/scripts/
or via *NetFreebies.Net* http://www.netfreebies.net/

A brief word on directory or folder structure

Your web site will not, of course, actually be a graceful edifice of linked pages and illustrations, it will only be an illusion of such, created inside a web browser and the user's head. What it will actually be is a series of files and directories or folders. Do yourself and your web server administrator a favour and spend a moment planning out a manageable and easily maintained file structure now. Otherwise in three months' time, when you are trying to find that stunning photograph of a tie beam, not only will you not be able to find it amongst the hundreds of files in your site, but you will not even remember what you called it.

File naming conventions

Call your HTML, graphics and other files something recognizable, and keep the names consistent. Even if your computer's operating system allows for spaces in file names, do not use them; likewise, stick to letters, numbers and dashes [-] or underscores [_]. This will ensure that your file names are recognizable to other people's web browsers. Give the files extensions (like '.html') if your system does not do this automatically. Finally, remember that the name of a file may well be part of the URL, or web address, that students need to remember in order to go to a particular page—so keep filenames reasonably short and easy to remember.

Folder or directory naming

Use a similar naming convention for the folders or directories in your site, and keep the names simpler and shorter if you can.

Table 3.3 Common file naming conventions

Description	Common naming conventions
Folder containing the site, a study companion for a module called FHY109	fhy109
home page of the site	fhy109.html fhy109.htm contents.html (avoid "index.html", as some servers will interpret other instructions for this name)
Folder for all the site's graphics	pics images
Graphic title header for the home page	main_title.gif MainTitle.gif main-title.gif
Introduction to subsection about war and society	war.html WarIntro.html war_intro.html war-intro.html
First page in a small linear run of pages	war1.html war_1.html war-1.html

Folder and file organization

The folder or directory structure of your site will determine the URL, or address, of any given page:

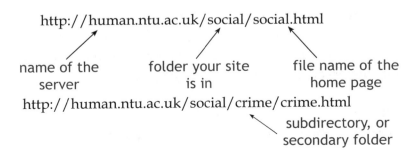

http://human.ntu.ac.uk/social/social.html

name of the server   folder your site is in   file name of the home page

http://human.ntu.ac.uk/social/crime/crime.html

subdirectory, or secondary folder

Organizing your files into folders or subdirectories is particularly useful if your site will be extensive, and allows for orderly expansion later on. Most sites at least have a folder in which to store all the graphics files together. What other directory structure you decide upon will depend on your material, but try to strike a balance between ease of maintenance (= more folders) and convenience of web address (= fewer folders). Many sites achieve this balance by placing 'top level' or frequently accessed pages in the main directory, then storing away in sub-folders the pages that people will rarely need to access direct.

> **Main folder/directory**
>> Home page
>> Site map
>> Top-level section menus
> **Folder for graphics**
>> Graphics common to all pages
>> Title header graphics
>> Graphics for top-level pages
>> **Folder for 'section one' photos**
>>> Photos for 'section one'
> **Folder for 'section one' pages**
>> Pages for this section
> etc.

Default page

Many web servers allow you to nominate a page that will load by default when the address for a certain folder is called up by a browser. This allows you to offer a shorter (and therefore easier to remember) address for your front page:

http://human.ntu.ac.uk/social
instead of   http://human.ntu.ac.uk/social/social.html

…where "social.html" has been defined as the default page. Ask your web server administrator or Internet Service Provider (ISP) whether this is possible on the server that will host your site, and how to do it. It may very well be as simple as making a copy of, short cut to, or alias of your front page and calling it "default.html".

Visual (and aural) design

It is imprudent to ignore the implications of the visual impact of objects, whether consciously designed or not, whether a typed memo, a car, or a multimedia project. The overall look and feel of a web site is, of course, an element of its design, and a very important influence on how the students interact with the site. The approach to visual design of web sites is also a topic of fervent debate. In essence, the polarized positions of this debate are taken by those for whom consideration of visual design is at best irrelevant and at worst a cumbersome distraction, at one pole, and by those for whom aesthetic principles outweigh other considerations, at the other. The reasons for these standpoints are numerous. A person who has spent a long time researching and writing some material may feel that any visual design considerations are entirely frivolous. (Nevertheless, this same person makes a design decision every time he or she presses the paragraph return key.) Or perhaps someone who has a sound appreciation of the technical constraints of web publishing may consider the look of a site to be much less important than its efficient operation. At the other pole, someone with a keen appreciation of aesthetics may consider the form to be much more important than the content. Or perhaps a person who has mastered a new and difficult technology may wish to showcase the fruits of his or her labours.

This argument is not a case of 'decoration' or 'no decoration'. There is an obvious difference between deliberate minimalism and a lack of consideration of visual impact. And a visually crowded page, with many graphic elements, is not necessarily a well-designed one. It is a question of balance. Careful consideration of the visual design of your site will balance your aims, the needs of the users and any technical constraints. For example, given that we are considering the development of a site for learning, certain assumptions can be made. On the one hand, it is probably reasonable to assume that your students may appreciate beauty in your pages; on the other hand, they will very probably become frustrated if functionality is subsumed to aesthetics, or if the beautiful graphics take half an hour to download. They will probably find endless pages that look very much like long paper documents a less than stimulating environment for learning, even if they appreciate the quality of the material and the rapid speed of access. How far these factors are true for you depends very much on your students and your material — in fact, on the planning decisions you made at the outset. Again, it is a question of balance.

Ingenious visual design can make content easier to read and follow, aid a sense of orientation and increase the student's enjoyment of the whole experience. A site that is a delight to use and to look at is a site to which students will feel encouraged to return.

Any decisions you have made about the navigation and organizational metaphors will influence (or reflect) the general look and feel of the site. A visual theme not only lends consistency, but, if chosen with care, can also transfer new perceptions on to your material. For example, if you suspect that your students perceive a particular topic to be dull, or a necessary evil, visual design that borrows from a more glamorous milieu may well lend it some of its charm. It may sound insincere—it also works. But however useful a theme may be, do not stick to it too rigidly, or it will risk becoming overt and tedious.

> William of Baskerville:
> 'How beautiful the world would be if there were a procedure for moving through labyrinths.'
>
> Eco 1983: 178

The rest of this section deals with some particular visual design issues, namely, page layout, readability, accessibility and technical constraints. Graphic elements themselves will be discussed in Chapter 6. There are, of course, other issues of visual design in web sites, and new considerations develop all the time. In light of this the best advice we can give is to stick to your plan, to be on the lookout for good ideas in other sites and not to follow new technologies heedlessly.

Broadly speaking, there are two groups of visual design elements to consider—those that pertain to many media (paper and electronic) and those that are particular to the web. Effective web design will aid readability and access as well as accounting for technical constraints.

The overriding factor in evaluating the design of your web site is not whether it is stunningly elegant, but whether it attains your objectives. Can the students find their way around the site? Are they motivated to use the material? Does the site work?

Organization at page level

As part of the process of making an effective, multiple-page paper document—perhaps an advert, a newsletter, or a book—it is common practice first to build a grid that will guide the layout. This grid charts elements that will repeat on different pages (like headers and footers) and how the content will be arranged. The grid is built by using horizontal and vertical lines to divide the page up into columns and boxes. This outline then influences where elements can be placed, so although each page in a document may look slightly different, a grid gives some consistency to the layout. It provides the reader with visual clues as to the relative importance of different items and as to what to read or look at next. How do you decide what to read in a newspaper or magazine? While you are reading an article, how do you know where it continues? Where it stops? The grid layout aids the reader in navigating pages and material. A grid might be overt, like that in a multi-column newsletter, or it might be less discernible, but it will be there. It might be followed rigidly — where, for example, photographs always appear in the same place — or it might be used as the basis for a more dynamic design, where text and graphics are interchanged on different pages. A grid approach works very well for web pages, too — giving a consistency of layout with some flexibility, and contributing towards a sense of unity across the site. Ideas can be developed from the grids used in paper productions, with the proviso that multiple columns of text will hamper reading if the student has to scroll downwards at all.

It is common practice to begin experimental mock-ups for a web page with paper and pencil, or with a desktop publishing or drawing package, before any actual web page writing is involved. You can create a grid for your site (Figure 3.1) without knowing any HTML. This way, your design ideas may be evolved without undue restrictions from your current level of technical knowledge. Sketch out what you would like to be able to do, not what you currently know how to — within reason. It is useful to note that HTML was never intended to be a tool for page layout — so you may have to modify your design a little, later on.

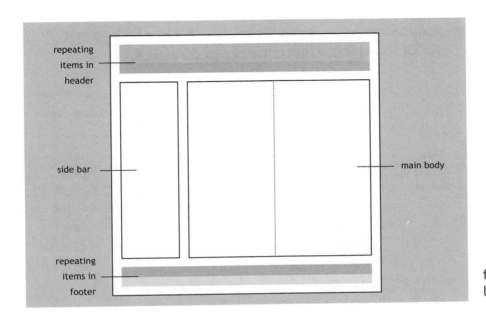

repeating items in header

side bar

main body

repeating items in footer

figure 3.1
layout grid

As you are sketching out ideas, build in the elements that will appear on each page—such as running headers and footers or navigation. Group like with like and always keep your objectives and the type of material to the fore. A useful layout for pages with little text, but many graphics, will probably not suit text-heavy pages, and vice versa. For inspiration, have a look at the page layouts of other web sites, educational or commercial, and adapt ideas gained to your own ends. (Adapt, not imitate, or you will, of course, incur the wrath of the site's owners.) Often-used styles include:

- Two- and three-column 'newsletter' layouts, for newsy sites, or those presenting text-based articles.
- Two-column, text and 'menu' side-bar layouts are used for a wide variety of sites.
- One-column 'tower' of text, for commentaries, rants and monologues—particularly those that are updated frequently.

- Plain, one-column pages, where there is a lot of text to read, or which are intended for printing.
- Single-screen pages, where concept is more conspicuous than text, where it is desirable that the visitor see everything on the page, or where the content of the site is broken down into 'bite-size' chunks.

Some general advice for page layout design

Keep it simple. The most effective web pages are often those that achieve their end in the simplest way. Avoid unnecessary visual clutter—if you are not sure why you are including a particular element (say, a picture or an animation), it is probably not needed.

Flexibility. Varying your design slightly for different areas of a large site will help the students orient themselves and will stop the site from becoming visually monotonous. For example, you could keep a consistent grid layout for all the pages, but use different colours for different groups of pages.

Essential elements that should appear on every page are the date the page was updated, a link to the site's main page and your email address or some means of contacting you. You should also append text links as an alternative to any graphic navigation. Include citation references, particularly if the document is likely to be printed and read later—this might include the URL, or title and page number. (You could complement this with a page giving help to the students in citing web sources.) State or link to copyright statements, if you feel they are applicable. These essential and common elements usually form a footer to every page and often appear in smaller text.

Making the page 'title' meaningful is very helpful to the student.

The title is the text that will appear on the bar at the top of the browser window. It is also the default name the browser gives to the page if it is added to the bookmark or favourites list. Hence 'FHY109: Medieval history: war and society: seminar 1' is more useful as an aid to finding the page again than 'page 12'.

Put important items at the top of the page. If your pages are going to be deeper than the browser window can display in one screen, remember that people do not really like to scroll. Put the most important information at the top of the page, including the navigation (which you should also repeat at the base of the page in some form). This is why banner advertising almost always appears right at the top of web pages—many people may not see anything further down. Never relegate anything really vital to the lower part of the page.

Many sites use intra-page navigation on really long pages. For example, a table of contents at the top links to information further down the page, where a 'back' or 'return to top' link skips back to the contents. Page authors like this technique because it is straightforward to create and labour-saving. Users hate it because they find it disorienting. Even so, it is polite to put in some kind of 'return to top of page' link in a very long document, as it saves the effort of dragging the scroll bar back up.

Keep headers and footers consistent, so that the student's eye can quickly pick out (particularly navigational) information.

So **passé.** In any fashion, after something has been very popular it becomes dated very quickly. Web fads change even more quickly than terrestrial ones and at the time of writing, it is still too soon for a retro craze for <BLINK> tags. Save yourself from a potential contretemps by avoiding the more extreme of new web crazes and the more outdated techniques. For example, frequent web users consider it more elegant to employ meaningful words for text links—that is, 'Download the <u>module booklist</u>', rather than, '<u>Click here</u>'. Similarly, avoid any initial phrases that include the words 'welcome to the…home page' and beware of extravagant but functionless introductory pages. On the other hand, if you like it and it works, use it. There will always be someone willing to offer unconstructive criticism of your site's design, so please ignore them.

Reading and readability

Providing information is one thing—persuading people to read it is quite another. If you plan to incorporate any more than the minimum of text into your site, you will need to present it in the most readable form possible. It really is the minority of people who will read everything put in front of them. For example, in reading a magazine or brochure, most people will read the headlines and probably the captions under the pictures, too; many will also read some of the leading paragraphs. Very few people will read most of the articles all the way through. As far as your students are concerned, independently electing to read certain information may be just what you would like them to do—but they should make this selection on the basis of their learning needs, not on how off-putting the text looks to wade through. Research conducted by Colin Wheildon in Australia found that a change of layout style of a page could increase readership from 32 per cent to 67 per cent and changes in the typography of the headline alone could increase readership by 38 per cent (cited in Parker 1995: 10–12).

Some of the techniques that make for readability and attract readers are the same for web pages as they are for conventional pages and some are peculiar to the medium.

Readability

- **Typeface.** In a paper document, you would usually promote readability by using a serif typeface (like Times) for the body of the document, reserving sans-serif typefaces (like Helvetica) for headings. For documents that will display on-screen, however, the "x-height" of a typeface like Times is too small for comfortable reading, unless the type size is set significantly larger than usual. A sans-serif typeface is, in fact, easier to read on-screen. Most sites use the widely available Arial. As in a paper publication, do not use more than two typefaces on a page, and use them consistently. And, of course, do not set the type in an extravagantly large or small size.

- **Font and case.** Italics are difficult to read, so avoid using them for long passages of text. Similarly, long passages of emboldened text can make for heavy going. Many people find placing text in upper case irresistible, especially when the information is important. But words in capitals, too, are difficult to read. The eye is trained to recognize the shapes of words while reading, rather than individual letters — placing text in upper case robs it of its ascenders and descenders and makes pattern recognition more difficult. Additionally, in online expression, upper case words are considered to have the effect of shouting. A type choice to avoid completely is that of underlining. A remnant of the days of reliance on typewriters, underlining not only visually cuts up words, but also in a web page may be confused with a hyperlink.

- **Justification.** Left-justified blocks of text (also called ragged right) make for the most comfortable reading, as the eye has a clear starting point for each line. Limit the use of right justification to small amounts of text.

- **Line length.** Both very long and very short lines of text make reading uncomfortable, so keep line length down, but not too far! Many page authors regulate line length by placing text inside invisible, fixed-width tables.

- **Contrast.** Make sure there is a high contrast between the colour you choose for the text and the colour you select for your background. If there is a lot to read, black text on a white background works best. Its reverse can be striking, but is also more difficult to read. For smaller amounts of text, or where reading is not essential, more daring colour combinations could be used. Similarly, if you decide to use a patterned background, choose a very subtle pattern.

- **Distractions.** Constant movement in the field of vision is, not surprisingly, a distraction that makes it very difficult to concentrate on reading. Hence animated text and graphics must be used judiciously. For example, the single line of 'ticker-tape' horizontally scrolling text (which is already past the peak of its popularity at time of writing) can make reading particularly uncomfortable. It generally moves right to left on the page or the status bar of the browser — in the direction opposite to the motion of an eye reading a text in a Western alphabet. The <BLINK> tag, popular around 1995–96, makes text flash on and off and was so reviled that whole sites appeared dedicated to its prohibition. If you would like to use animation on your site, first consider how suitable it will be given your aims, then take care in its use. Be very wary of placing continuously animating objects near to any significant amount of text. Alternatives could be to have the animation cycle once, or infrequently, or to allow the user to switch it on and off. The latter options works well for animations that are needed to illustrate a point described in your text.

Motivation to read

- **White space.** Leaving empty space attracts the attention to, and aggrandizes, the objects elsewhere on the page. It can also be used to give the illusion that there is less text to read than there really is, making the page look less daunting for the reader to tackle.

- **Colour.** For many people, writing a web page is the first time they have been able to make full use of colour in a publication. Those of us with limited budgets have seldom been able to have a paper publication printed in two colours, never mind 216. But as with so many things in life, so with colour—just because you can, does not mean you should. Selecting a limited palette and sticking with it (photographs excepted) will make sure your students are not distressed or distracted by an overly boisterous colour scheme.

- **Break up the text.** A text-heavy page looks less intimidating if it is broken up into small blocks. **Sub-headings** are the most obvious device for achieving this effect (make them conspicuous), but you can also use:

captions	brief descriptions or comments under pictures, or in a scholar's margin
pull quotes	occasional extracts from the text, written large and in inverted commas

lead-in paragraphs	frequently used in newspapers and magazines, this technique places the text of the first paragraph in larger type
side-bars	a short piece of text running at the side of the 'main' body of text; perhaps a comment on information related to the overall topic
boxes	information of particular interest is highlighted by being placed in a plain or tinted box
rules	horizontal or vertical lines used to break up different text segments, columns, or changes of topic
bullets	bulleted lists offer an easy connection with the text

- **Graphic devices.** If you place a prominent graphic of an exclamation mark beside a piece of text, most people will read it. Symbols that have a generally understood meaning can be used to draw your students towards important information that you do not want them to miss. Text set in large quotation marks, for example, is almost irresistible to read. Take care that the reference you are making is not specific to one particular culture, but is widely recognized.

- **Numbering.** Numbering, like bulleting, draws attention to the significance of a list. People are likely to progress through a numbered list, not skip over it. For this reason, numbered points are used in promotional material when copy deadlines are tight. Numbering and bullets are to text-heavy pages what overheads are to lectures.

- **Viewing patterns.** Consider how your students' eyes will move over the web page—they will probably start near the top, slightly left of centre—and think about ways to lead them into the text.

- **Less really is more.** No matter how creatively you design your pages, if there are too many elements and too much text on the page, the students will find it difficult to commit to reading them. Simplify the pages, cut out or move some text, be consistent in the use of any graphic elements and use them sparingly. And leave plenty of white space.

Accessibility

Organizing and designing your site well will improve your students' experience of the material, but there are one or two further steps to widen the site's accessibility. Widening accessibility here means, in practice, considering those of your students who are blind, partially sighted, dyslexic, colour blind or deaf. In consideration of those who are colour blind, make sure you provide alternatives for information that is colour-dependent in its presentation. If you are using audio material, or video with an audio element, provide text transcriptions, captions, or some equivalent representation. Students who are blind or dyslexic may well be using software that reads the text aloud, so there are several features you can use in your pages to make this process as smooth as possible.

- **Text only.** The clearest way to improve accessibility is to provide a parallel text-only page for every page on your site. However, this may not always be possible, due to your time constraints.

- **Make full use of the HTML tags.** Rather than being principally a layout tool, the primary function of HTML is to define the different elements in a page. Many people never use the tags for addresses, quotations, table captions and headers, because the same visual effect can be obtained with the tags for italics or ordinary table cells. But if you are designing with a text reader in mind, defining all the elements accurately can only add to the clarity of your material. Some practical advice for writing HTML is included in Chapter 7.

'If you can't figure out any other way to make a page accessible, construct an alternate version of the page which is accessible and has the same content.'

CENTER FOR APPLIED SPECIAL TECHNOLOGY (1999)

- Give text alternatives for graphics. The tag for graphics can take an attribute called "ALT", which allows you to include a text label for each picture. Thus students using screen readers and students using their browsers with the graphics 'turned off' can tell what the pictures represent. Where longer descriptions are needed, for example for a photograph or a graph, you should provide a separate text description, too. At time of writing, the current generation of browsers do not acknowledge the <LONGDESC> tag for graphics. The current convention is to use a "d-link" — a letter D placed beside the picture, which acts as a link to another page containing a full description.

- Provide single-column alternatives for text placed in multiple columns.

These are a few examples of techniques for improving accessibility; more exist and more are developing all the time. The web sites of the Center for Applied Special Technology, the National Center for Accessible Material and the World Wide Web Consortium offer invaluable advice on this subject. The former also has an online page validator, *Bobby*, that will check your pages for accessibility issues.

Technical constraints

The technical constraints on the interface design of your web site comprise not what it is currently possible to achieve, but the abilities and limitations of the technology to which your students have access. You have probably already considered this as part of the planning procedure. Designing with technical constraints in mind is again a question of balance—that of higher functionality with wider usability. As a general guide, the newer the technology you use, the more likely that you will limit the number of students who can use your site or see it in all its glory—those with older browsers, slower connections or older computers. Below we offer a summary of some of the technical issues at the time of writing, and their implications for the design of your site.

> 'The "eye candy" effects of animation and flashy graphics often mask both a lack of content and an incoherent maze of links which any discerning visitor is glad to leave quickly via the nearest exit.'
>
> JOHNSON 1998

Speed

How long do you wait for a web page to download? Twenty seconds? Measure it; 20 seconds is a long time. How long a page takes to finish loading in someone's browser depends on many factors—the speed of their modem or network connection, the amount of traffic connecting to the server, the size of the files being downloaded. It is unlikely that you will be able to do anything about bandwidth, but you can consider the file sizes of the pages and graphics in your site.

Large, complicated graphics, animation effects, JavaScript, numerous hyperlinks and longer pages will all add mass to the total amount of data being transferred for a given site. So while you are designing your site, the trick is to balance the content with an appreciation for the speed of download. There are numerous 'page checkers' on the web, like *Bobby* (discussed in 'Accessibility', above) that will assess the combined file sizes of a page and its graphics and calculate how long it will take to download over a medium-specification modem. It is a sobering process.

You can achieve balance for your pages by a combination of compromise and communication. If you need to include large files for your students to download (say, a version of the module handbook that they can open in a word processor) then let them know what it is and how big the file is before they commit to downloading it. For example:

FGG221 Social geography handbook. Available as:

> Web pages, 5
> Pdf, size 33k
> MS Word document, size 130k

As far as a page itself is concerned, it will load more quickly if it is scripted neatly. Browsers will accommodate less-than-accurate HTML, but will load the page faster if, for example, the sizes of graphics and table cells are defined by you in your script. (This is described in more depth in Chapter 7.) Fewer hyperlinks also equal a faster page, so limit these to the links you really need for navigation, and the most useful links to other sites.

Techniques for making efficient web graphics will be discussed in some detail in Chapter 6, but there are a few general guidelines to graphics and file size to bear in mind:

- A graphic only has to be downloaded once per session, no matter how many times you use it. It makes sense, therefore, to use elements like headers and buttons over and over.

- Use graphics that are physically small and have fewer colours.

- Give your students a choice whether to view a photograph or not — include a tiny version of the photo (a thumbnail) that links to the full-size picture and indicate its file size.

As a rough guide, the consensus between most of the web developers that we know is that you should aim to keep each of your pages, with its attendant graphics, to a total size below 20k.

Browser degradability

Degradability or backwards compatibility means consideration of those of your students who, for one reason or another, may be viewing your site with browsers older than the current version. If this is likely, you should investigate whether the latest scripting developments will work in older browsers before you use them. (Even current versions of different companies' browsers have different capabilities.) It is advisable to provide alternative options to any generation-specific feature in any case, most particularly if it will not work in the generation of browser immediately prior to the latest one. Many people do not upgrade to the latest version of a piece of software immediately, and this is often an active choice. Your students may well not have control over which browser they are using. So, if the piece of HTML or JavaScript you want to use does not work on the majority of browsers to which your students have access, leave it out. If it works on 80 per cent of them, use it, but provide an alternative that does not spoil the students' experience of your site. Fortunately, finding out what browsers can and cannot use is usually straightforward. For example, at their developers' site, the Netscape Communications Corporation (1998) indicates which version of JavaScript will work with which generation of their browser.

Control

You have much less control over the appearance of a web publication than over a paper publication. Unless they are working in a computer lab where they are locked out of the browser's controls, the students will be able to alter the look of your pages by changing the typeface and size in which the text displays, changing the background colour and choosing to 'switch off' the graphics. You can only accept this and work with it — make your page layout flexible enough to function well, even with these changes.

Platform

Although web technology is not essentially platform-dependent, there are differences that influence the way your pages appear. Are your students likely to be using your site from different platforms? The two most widespread at time of writing are the Windows and MacOS platforms and they do display pages differently. Typefaces tend to appear much smaller on Apple systems (or much larger on PCs, depending on your point of view). They do not share quite the same set of system colours, effectively limiting your choice from 256 to 216. Downloadable files you provide may not have readers or clients that work on both systems, although this is becoming less likely as compatibility between the two systems improves. Any typeface you use in your text (not in graphics) must also be available on the recipient's machine, or it will return a default (usually Times) — which is why so many sites use the widely available Arial.

Hardware

The different specifications of the computer hardware to which your students have access will also affect the way your pages display. For example, although at the time of writing most designers work using 17 inch monitors running at a resolution of

1024 × 768 pixels, most people using computers at home use 15 or even 14 inch monitors, running at a resolution of 800 × 600 pixels. So a page that fits perfectly into the browser window on one machine will need to be scrolled on another. This is particularly important with regard to page width, as sideways scrolling is probably less instinctive than downwards scrolling – your students may miss some information altogether if the page overlaps the browser window slightly. A page width of 600 pixels should work with both resolutions and sizes. If you wish to make the whole page fit in a browser widow, make the depth about 450 to 500 pixels. If you know that all your students will have monitors of a certain specification, then you can, of course, design with them in mind.

Plug-ins

Viewing material made with the latest technologies often requires plug-ins or software extensions extra to those that were released with the browser. The pattern of acquisition is theoretically that people try to download an animation, or a graphic, or an audio file, are advised that they do not have the right plug-in, then go to the parent site to download it. Or perhaps they register at the parent site, and their software updates itself automatically from time to time. In practice, people may defer downloading new plug-ins, or never actively acquire them at all.

You and your students may have no control over what plug-ins are loaded on to the machines they use, if they are in a public library or an institutional computer room. This can be frustrating if a new technology provides just what you need to deliver your material in the most apposite way (see Chapter 5). Unless you can persuade all the relevant parties to install the plug-in, you will have to find another way to achieve the same effect. There usually is another way.

Further reading

Barron, A.E., Tompkins, B. and Tai, D. (1996) Design guidelines for the World Wide Web, *Journal of Interactive Instruction Development*, **8**(3), 13–17

Center for Applied Special Technology (CAST) (1999) *Bobby 3.1.1* http://www.cast.org/bobby/ (6.9.99 edition)

Cleland, J.K. (1995) *How to Create High-Impact Designs*, CareerTrack, Boulder, CO

Jones, M.G. and Farquhar, J. D. (1997) User interface design for web-based instruction, in B.H. Khan (ed.) *Web-Based Instruction*, Educational Technology Publications, Englewood Cliffs, NJ, pp. 239–254

The National Center for Accessible Material web site has useful advice for practical ways to make web pages more accessible. http://www.wgbh.org/wgbh/pages/ncam/ (15.3.99)

Shiple, J. and Yi, Y. (1999) The death-to-download ratio, *Webmonkey*, 30 August, http://www.hotwired.com/webmonkey/99/36/index0a.html

Veen, Jeffrey (1998) Bad bandwidth, good design, *Webmonkey*, 19 October, http://www.hotwired.com/webmonkey/98/42/index0a.html

Webmonkey has a very good collection of other articles on web design and redesign. http://www.hotwired.com/webmonkey/

World Wide Web Consortium, Web Content Guidelines Working
    Group (1999) *Web Content Accessibility Guidelines*
    http://www.w3.org/wai/gl/ (5.5.99 edition)

chapter four
building successful content

The medium is the message

Marshall McLuhan *The Gutenberg Galaxy* (1962)

D esigning your own whizz bang home page for fun is quite different from constructing a series of pages into a successful package which will engage, stimulate and inform students. As we have seen, the key to successful web page construction is planning, planning and more planning. This relates not only to the technicalities of information architecture and navigation which form the skeleton over which your material will be draped, but to the material itself—pointless having fine bone structure but saggy flesh and bad skin. Considerable thought must go into selecting your material. The Stygian and esoteric topic of copyright—your own and that of others—must be addressed. You must consider the type of media you might want to use and, more importantly, the type of media it is feasible to use. You must formulate yet another plan, at this stage concerning the collection and putting together of the material. It is also important at this juncture to consider assessment. Here, attention should be given to the procedures, criteria and methodologies you might want to use in assessing the students *and* the package you put together (assessment is discussed in more detail in Chapter 9).

Selecting material

At the heart of any pedagogical process is the selection of material which the teacher wishes the student to access, assimilate, digest, dispute, challenge, consider, question, review, etc. When creating a module or course, teachers normally (students and colleagues may have other opinions) seek to provide information, often in a variety of formats, which will best convey those aspects of the subject area which they consider to be most appropriate and intelligible, given the nature of the discipline and the academic level of the student. Generally speaking, the same aims apply in selecting material for web-based instruction (WBI). However, there is a need to 'think on'. Thus far, most of the instructional materials available on the web have been informational. Most of us encourage our students in the critical use of material they find when surfing the web. Many put curricula, syllabi, handouts, assignments, past examination papers and lectures notes on the web. But there must be more to WBI than simply posting analogue materials on web pages. There is, as Sugrue and Kobus (1997: 38) point out, 'more to instruction than information'. Thus, while the nature of the material you are using in your WBI may be similar to that used in your more traditional teaching, there is a need to be adventurous, innovative and experimental. Indeed, as Spira (1998) notes, 'progressive ideals should be considered whenever technology is used to enhance the learning environment'.

> 'Simply providing well-designed curriculum and interaction opportunities is not enough to ensure students' success in a Web-based learning environment.'
>
> McVay 1998

Presentation methods have come a long way from the back-to-the-class nose-in-the-chalkboard scenario of the not too distant past, and very proximate present. Yes, you too will know of colleagues who are still at that Palaeolithic stage which leaves technoliterati in the class staring at each other in disbelief the first time they see it in university—in these circumstances Morpheus is never far away. If you do not actually know such creatures personally you will see evidence of their activities—unintelligible hieroglyphs on the chalkboard, or (often with indelible pen) on the (for them) newfangled wipeboard; projection screen wound up in such a way that it causes considerable effort on your part (to say nothing of the great amusement for the class) to get it back to where it should be; overhead projector dumped in a corner somewhere; lectern dumped in another; and so on. As Kearsley (1998: 49) remarked: 'One of the saddest aspects of educational technology is how ill-prepared most teachers are to use it — despite widespread attention to this issue.' Indeed, it is the mindset of teachers and administrators that can be the biggest obstacle to effective adoption and implementation of WBI (Gray 1997).

Yet, audio-visual aids have progressed considerably over the past 20 years or so. For some time, we have been able to use slide or overhead projectors to bring colour into our presentations, either as a way of enhancing the intelligibility of an image, as in a map or photograph, or simply to brighten up an otherwise dull graphic. This ability to include previously prepared text and graphics in a range of colours has been greatly advanced by the development of Microsoft *PowerPoint*. With a relatively low skill base and minimum effort it is possible to create an effective and arresting display incorporating all the elements just described with the added advantage of drive-in, flying, drop-in, typewriter, laser text and other eye-catching features. This medium generally precludes the transportation of piles of lecture notes and slides, and your entire presentation, module or course can be conveyed via a pocket-size floppy disk or, better still, accessed via a passworded directory on a server. Notwithstanding the problems of hardware provision and Murphy's law, this presentation package is now widely used in some UK higher education institutions and has become an established part of the instructional repertoire of the US faculty (for example, Klonoski 1997; Ptaszynski 1997a; Sammons 1997; Sommer and Anderson 1997).

Many courses, not just media studies, now make use of bespoke or other video or audio analogue or digital material. These form an excellent supplement to conventional lectures and seminars. The great beauty of WBI is that all of these can be combined in new and interesting ways which, coupled with interactivity, can create an exciting and multifarious vehicle for your material. An added bonus—some might argue the main attraction of internet delivery—is that the material can be accessed in an open manner and be delivered at a distance.

As with more conventional methods of learning and teaching, choice of material will depend on a number of criteria. Some of these will be generic to most disciplines and some will be subject specific. Here we use examples from our own WBI packages *Social Geography and Nottingham*, *e:net* and *Medieval History*. The first is fully integrated into a second year geography module and has been running since 1996. The second is a supplement to a third year international relations module and has been running since 1998. The third project is designed as a support for first year students and ran for the first time in 1999.

> 'Sure, the Web talks a good game with its sound and video and animation and god-awful 3-D interfaces. But lurking beneath all those various bells and whistles is good ol' text. It doesn't have the sinus blowing sex appeal of Flash or MP3, but text is the stalwart backbone of Web-based content. It rolls up its sleeves and gets the real work done.'
>
> ALLEN 1999

Despite the scope for multimedia in web pages, text is still the main medium regardless of the efforts of the corporate web designers and elements within the advertising industry who would argue otherwise. Text is often the most overlooked and understated medium used on the web, yet can be the most powerful and significant, if used properly (Montgomery 1998). The value of text and the need to incorporate good quality text into web pages is often forgotten as designers become overawed by the multimedia available to them. Substance takes second place to image. 'Since writing is the lowest common denominator online, it often gets bottom priority when a site is being put together' (Allen 1999). That said, text on web pages requires careful consideration and more thought than with other, more conventional media

Untangled web: developing teaching on the internet

(O'Carroll 1997). Although we would all give careful consideration to our undergraduate handouts and research papers in terms of content and presentation, writing on the web requires additional planning. Most of us are not yet used to reading at length from a computer screen. Light fired onto glass does not have, and probably will never have, the appeal of paper. As Sophie Vandebroek, vice-president of Xerox research and technology has conceded, 'people will always want the feel and smell of paper' (cited in Veash 1999: 5). Most people, as Joshua Allen (1999) points out, 'can only stomach small doses of online reading'. It is important, therefore, to pay particular attention to text selection and text construction when designing online learning and teaching materials. Points to consider are:

- How is the balance to be struck between the amount of material you want to convey and the amount the student will sit and read before clicking on? Even a page full of text can induce pixel-blindness and lead to page-hopping.
- What function does the text serve? Is it integral to the prime instructional aims? Is it part of the navigational structure? Is it supplementary material?
- Make full use of different font sizes, types and colours to distinguish these without making the overall appearance too noisy (see Chapter 3). Whatever system you adopt, be consistent from page to page.
- Divide your text into manageable chunks.
- Avoid situations which involve too much scrolling. In our experience students are not overly fond of scrolling.

As we have mentioned earlier, there are technical and design reasons why over-use of flashy graphics and gizmos should be avoided. From a pedagogical perspective pages which are reliant on memory-intensive objects should also be avoided. Clutter and excessive elements should also be avoided when planning and constructing pages (Kaur *et al.* 1999). This is because:

- Slow download leads to distraction and concentration lapse.
- Badly planned and excessive graphics lead to noise and concentration lapse.
- Poorly linked and structured pages waste time and cause frustration.
- The message must not be subordinate to the medium.

> On one WBI 'there is an animated mouse who appears to be on cocaine or has eaten too much rat poison. At first glance he/she is amusing but, at the second or third look, is intensely irritating.'
>
> BOSHIER *ET AL.* 1997: 347

As the next section shows, it is desirable, as far as possible, to use as much material generated by yourself, or sympathetic colleagues, in your web pages as possible. Write your own text, take your own photographs, make your own video clips, draw your own diagrams, etc. This may seem like a long-winded way to go about things when something you already require exists in a perfectly presentable format. However, you may well find that the time and effort required to get copyright clearances for many materials mean that it would have been easier, quicker and cheaper to have generated your own in the first place. Copyright owners, who would normally not hesitate to give clearance for their work to be copied onto conventional media, have a tendency to be more cautious about allowing their work to be incorporated into web pages. Very often, obtaining copyright permissions can be some of the most difficult aspects of WBI (Whalley 1995). Not all subject areas are in the fortunate position of being able to generate original material for WBI. For contemporary geography or sociology, for example, it is relatively easy to capture your own images to get a message across (Figure 4.1). But if you want to incorporate an historical dimension by looking at, say postwar housing, then searches through photo libraries might be required. This in itself will require time, but added to this time budget will be the search for ownership and the application for permissions. Even creating a medieval history WBI is not without permissions problems. Original documents from the period are well outwith the copyright dates, but even using these might require reproduction clearances. The same applies to the reproduction of plans of medieval towns. Nevertheless, it is possible to visit towns, such as York or Bruges, with tangible medieval remnants and make your own photographic or video images. Also, consider using copyright-free materials, many of which are increasingly being made available by web users frustrated by present conditions. Clipart can usually be used without permission (Browell 1997). There are also multimedia databases available under site licence which can save the WBI creator considerable time and expense (Morrison 1997a).

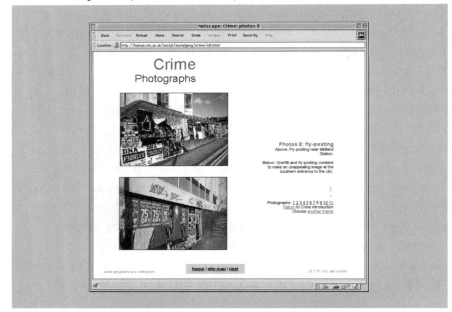

figure 4.1 use your own images to avoid copyright problems

Copyright—yours and others

There are a number of ways that copyright can be viewed: user–owner; fair use–piracy; monopoly–free access; protection–dissemination; infringement–licence; and so on. But copyright 'is essentially a system of property. The province of copyright is communication' (Strong 1994: 1). The property notion has logic and tangibility when we consider conventional media. As Mitchell (1995: 136) has noted, when you check a specific book or video out of a library then, for all intents and purposes, it is denied to others. 'By contrast, the digital resources that are available in cyberspace do not have to be scarce resources. And it is a queer kind of property that can be valuable without being intrinsically scarce.'

For most academics copyright is not a major issue. As Linda Enghagen (1998) puts it: 'Most of us will complete our careers in higher education without the whisper of a threat from litigation.' Yet, we like to keep within the laws as much as anybody can understand them. The most common infringement is photocopying when we may copy a larger percentage of a journal or book than is allowed or else make more copies for student handouts than the copyright authorities, rather than the authors, might be happy with. With the pressure of controversial research assessment exercises on many individuals and institutions there is an increasing temptation for some to borrow and incorporate the work of others in their own work without making the usual acknowledgements. At the moment this is not a major issue and is often resolved without recourse to the

'Copyright is about encouraging the creation of new works. But it is also about access to works by the public and about fair practice uses of such works in education, research, private study and reasonable non-commercial consumer copying within the home; about access to and fair practice use of such works through libraries and by disabled people; and about new technology which provides the means by which rightowners make their works available to consumers, as this provides access to culture.'

EFPICC 1999A

law courts. Although it has always gone on, it is likely to increase and thereby bring opprobrium on the academy. Since many of us are authors ourselves we tend to be aware of our own copyright but to be quite blasé about the whole issue, since most of us do not rely on royalties — which for academic writing are non-existent or relatively small — as our main source of income. Contributors to academic journals blithely sign away their copyright to the publishers. Contracts for academic books are generally on a sounder footing, but few academic writers are likely to lose vast quantities of sleep if a student or peer were to copy more than n per cent of the tome to further their studies.

But, as Negroponte (1995: 58) has noted, 'Copyright law is totally out of date. It is a Gutenberg artifact.' Most of our ideas and attitudes to copyright are based on laws formulated over three centuries ago (Zobel 1997). Yet, if anything, copyright laws are becoming more restrictive and anachronistic. With the web, the whole issue of copyright and intellectual property turns from the pain that it is in the analogue world into an unfathomable nightmare which the smartest minds in the international legal arena have yet to master. The fact that the web is a global medium, while copyright law varies between countries, adds to the complexity of the issues. There are a number of major differences, for example, between UK and US copyright laws, and even within the UK, the differences between Scots and English law add to the confusion (Charlesworth 1995). Significant differences exist between different European countries.

Within the European Union (EU), the current proposals to further tighten copyright law threaten to strangle the information age at birth. These restrictive proposals for manipulating material currently in copyright — particularly converting analogue information into digital format—are likely to turn us all into historians insofar as older, out of copyright material can be used without fear of sanction. The European Fair Practices in Copyright Campaign (EFPICC 1999b) is concerned that the EU 'Directive will allow powerful right holder corporations to undermine consumer freedoms by blocking personal digital copying, even when this does not infringe copyright'. There is a real fear that the EU proposals would alienate many 'from the benefits which the Information Society has to offer, creating Information Poverty for large segments of European society' (EFPICC 1999c). This is the paradox of the information age. We have the technology to manipulate and disseminate but we are often denied access to the information by legal or technical means. This is a problem throughout Europe and particularly in the UK. Politicians talk glibly about the information society and the information superhighway and have no concept of the real problems involved. We are some way, as Romiszowski (1997: 31) reminds us, from 'universal democratic access to the world's information resources'. Those of us who have colleagues who have academic interests in the United States can only watch in envy as they are able to browse the internet for information on a wide variety of subjects and then download and manipulate that information *free of charge*. The constitutional concept of the public domain has real meaning for American society. The American taxpayer has paid for census of population data, for example, and the American government and its agencies allow Americans, and the rest of us, to access that information free, or for a nominal charge. In the UK, and many other countries, the taxpayer pays for the collection of similar datasets, but the government then feels it has the right to sell the information back to the taxpayer. This mentality is hardly conducive to fostering an information-rich society. The new, so-called Freedom of Information Act looks set to make information even more difficult to come by in the UK. It is currently easier for UK researchers to use US archives—available under the real American freedom of information legislation—to find out about UK government and society than to try to access information at home.

A multimedia or WWW publication may include some or all of the following copyrightable components (Charlesworth 1995):

- Literary elements—protected as literary works.
- Dramatic elements—protected as dramatic works.
- Musical elements—protected as musical works.
- Artistic works (graphics, photographs, drawings and models)—protected as artistic works.
- Moving images—protected in the same way as films.
- Sound recordings—protected as sound recordings.
- Typographical arrangements of published editions of literary, dramatic or musical works.
- Computer programs—protected as literary works.
- Choreographic routines—protected as literary works.

Charlesworth (1995) goes on to give a worst case scenario: a web page dedicated to a biography of Margot Fonteyn, 'including mpeg excerpts from a film of Swan Lake, photographs of her performances and ballet music by several composers, living and dead, including pictures of the musical scores'. To create such a page, without infringing copyright law, it would be necessary to contact all performers (including the relatives or estates of deceased persons). And not just the featured performers would have to be contacted, but also backup musicians and singers, chorus girls and even crowd extras! In certain circumstances, the Copyright Tribunal, under the 1988 Copyrights, Designs and Patents Act, can give consent on behalf of untraceable or unreasonable rights holders — but only for the UK. Of course, even if it is possible to make contact with all necessary copyright holders, it needs just one instance of obstinacy to scupper all your plans. This — not software or hardware problems — is the reason why the web is not being used (and never will be, till the copyright mess is sorted out) to its full potential. Only the very large corporations can offer full multimedia encyclopaedic 'edutainment' products, given the time, energy and money required to clear copyright. Once more, the information age offers us a paradox. We are dealing, on the one hand, with a global, instantaneous and relatively cheap medium, while on the other, labouring under archaic and time-consuming clearance issues. We might spend a few days creating a set of pages which can be accessed anywhere within seconds, yet spend months, sometimes years, trying to find the owner of a few lines of text or an image. Then we have to convince them or their lawyers and/or accountants to allow us to make use of the material. Fortunately, there are a number of organizations which have amassed substantial libraries of material and have acquired the copyright for this. Thus, while clearance and fees might still be an issue, getting permissions is relatively straightforward with these streamlined organizations (Browell 1997). These one-stop shops are likely to increase as multimedia expands as an entertainment, instructional and pedagogical tool, via the web and otherwise.

There are a number of rights which a multimedia producer/WWW publisher might need to obtain under copyright law (Charlesworth 1995; Enghagen 1998). These include:

- The right to reproduce the copyrighted work.
- The right to issue copies of that work to the public and to let them copy it (limited).

Untangled web: developing teaching on the internet

- The right to adapt the work.
- The right to perform the work in public.
- The right to broadcast or publicly display that work.

As The Nottingham Trent University (1999) points out, copyright is automatic, and you should therefore consider everything on the Internet to be copyright, including:

- Ordinary text (e.g. .txt and .html files).
- Email.
- Photographs.
- Drawings.
- Graphics (e.g. .gif or .jpg files).
- Moving images (e.g. mpeg movies).
- Audio files (e.g. .au or .snd files).
- Programs (e.g. CGI programs).

Any uploading or downloading of information through online technologies that is not authorized by the copyright owner will be deemed to be an infringement of their rights. Infringements include:

- Scanning articles from magazines, books or newspapers to include on WWW pages.
- Duplication of textual material on your WWW pages where copyright belongs to another party.
- Downloading graphics from other WWW pages, or graphics archives, to include in your pages.
- Placing someone else's pages within a frame that does not show the real location but attempts to pass off the page as your own.
- Altering text or graphics without the consent of the owner of the copyright.

- Putting old examination pages on the WWW, which might contain copyrighted material within them.

When we consider copyright on the internet there is not just the matter of posting, say, an extract of text or data from another source — cleared or not. There is also the problem of breaching copyright laws by simply browsing the web and creating an automatic copy in the cache (Zobel 1997). However, caching by web browsers is automatic and not deliberate, and can be classed as fair trading in the UK (The Nottingham Trent University 1999).

> 'We are schizophrenic about the treatment of information in our society. Although we like to talk about living in an information age and an information society, we have yet to begin to comprehend the consequences of this shift, much less so to accommodate it.'
>
> BRANSCOMB 1994: 184

There is also the problem of including links in your pages. For example, the *Shetland News* site including links to copy in the *Shetland Times*, thereby incurring the wrath of the management of the latter newspaper, led to a much-publicized court case (Arlidge 1997). A user can breach the University of Minnesota's internet publishing policy, for example, by simply linking to a page that violates that policy (University of Minnesota 1998). One can only wonder what Joseph Heller would have made of all this had *Catch-22* been written about the late 1990s.

It would seem that just as big business has discovered and seeks to tame the internet, so the freedom, once seen as a key and distinguishing feature of the vehicle, is being jeopardized. For many years the domain of the military, academics and policy makers, the internet has lost its esoteric charm. It is no longer a renegade medium and has become mainstream. The pickings are too juicy for the corporations and lawyers to ignore. Too much litigation has already been spawned via the use of the web. Thus, many institutions have drafted their own guidelines for putting material on the web, including most universities. For example, The Nottingham Trent University issues an *Internet Code of Practice* which covers, *inter alia*, the Computer Misuse Act 1990, copyright, The Data Protection Act 1984, The Official Secrets Acts 1911–1989, defamation, obscenity, trademarks, discrimination and the institution's own Computer Users Code of Practice. These wider issues of libel, obscene materials, data protection, trademarks and software licensing are further legal considerations in creating web pages, over and above those of copyright (Charlesworth 1995; McCracken 1999; Naughton 1998; Thomas 1998; Westell 1999a; Wheatley 1996). The University of Minnesota's (1998) *Publishing Information on the World Wide Web* sets out strict guidelines as to the nature of some information which must be included in any web pages generated within the institution. It also points out that those creating web pages 'must have permission to publish the information, graphics, or photographs on their pages if the publisher is not the author or creator'. It would be good practice to check the policy of your own institution. If none exists, check with the institution's webmaster, if one exists. (See the Air Force Distance Learning Office (1998) for links relating to copyright law, though most of the sites listed relate to the US experience.)

As with other educational media, the use of copyrighted material on any WBI site should be cleared with the original copyright holder, if that is possible. At a time when corporate agglomeration apparently knows no bounds, in terms of scale, spread and speed, actually determining who controls the copyright of specific materials is becoming an increasingly difficult, expensive and arduous task. The following advice from Barron *et al.* (1996: 13) is worth bearing in mind: 'All copyright notices and release statements should appear in the footers of the pages. This will simplify the process for others who may seek to cite the materials or seek permission for their use.' This will lengthen your pages and bulk the content. It will make the pages noisier than they might otherwise be. But it is sound advice in these litigious times and should be adopted as best practice.

A further issue for academics is that while copyright ownership is usually with the creator, if the work has been created by an employee in the course of their employment, then copyright belongs to the University. This is often the case with other employers as well (Charlesworth 1994; McCracken 1999). As Westell (1999b: 15) points out: 'Under English law, there is no requirement to register a work in order to secure copyright protection. Neither is the addition of the © symbol a requirement for securing protection in this country although it is commonly used (and does no harm).' In other words, the act of creating an original work ensures copyright rests with you or your institution. Nevertheless, since your work could be accessed in countries with different laws, it is a good idea to think about protecting the copyright of your WBI in some way, even if by simply including a copyright statement and the © symbol. It might also be a good idea to consider password protecting your site. This will allow you some control over who accesses the material. Another security option is making the material available only on the local intranet. This gives less control over access than the password option and negates some of the objectives of open and distance learning.

Media

Generally speaking, apart from the nature of delivery, the main reason for using the web for ODL is the range of media which can be used and combined to create a more exciting and novel teaching and learning experience. When we think of the web we think, or should think, multimedia — or hypermedia. The National Council for Education Technology (1993) has suggested that multimedia consists of the use of two or more different forms of media — for example, tape/slide. Dahmer (1993) goes further and has defined multimedia as something that combines the capabilities of technologies that used to be separate. This can combine things like text, graphics, sounds and still or motion pictures in a smooth way to present training and information. One of the major problems with multimedia, related directly to the foregoing section, is the clearance issue. If most of the material you have amassed, and intend to use in your WBI, has been originated by you or sympathetic colleagues, contacts or friends, then you may find that only a small proportion of project time will involve chasing up clearances. If, however, you wish to use myriad sources over several media, and these are all held in copyright, then you must budget accordingly, in terms of both time and money.

Generally speaking, working with multimedia is more expensive than working in one medium. There are equipment costs to consider — should these be bought in, borrowed, begged? Do you have the expertise to use the equipment, to create your own images, sound, film, or whatever? Or could you again buy, or otherwise procure, these skills? If you decide to use existing material, do you have the expertise and equipment to edit and manipulate the material? And if you have these, or can acquire them, does the copyright clearance allow for such manipulation? You will find that textual material is not only the easiest physically to manipulate to suit your needs, but also the medium that is most likely to be allowed to be manipulated in copyright clearances.

It is worth bearing in mind, too, that clearance costs are typically around 50 per cent higher for the use of digital media than print. Clearing individual items from many sources can become costly. There is the administrative burden to account for, as well as the actual fees that may be incurred. When clearing multimedia, total clearance costs can escalate rapidly. It is important to maintain precise records of what material has been cleared and under what conditions. As McCracken (1999) has noted: 'This is essential if one is to be able to re-visit the product in order to extend initial clearance levels for further exploitation or to re-version.' All of these issues must be dealt with at the outset, and incorporated into your structure plan, time management plan and your budget (if one exists).

There has been considerable pedagogical debate about the way media influence learning, particularly in distance education. The question Carter (1996: 31) poses is worthy of consideration: 'Do media merely deliver content, or are they capable of influencing learning?' Although we have not had the opportunity to investigate this ourselves in detail, we would tend to agree with the conclusion she came to in addressing this question. Like any learning situation, individuals have different experiences and expectations of multimedia. Some find they can absorb information more readily when more than one medium is involved. In general, the feedback from students has been positively in favour of multimedia. Avoid, however, media overload. Some users may find noisy, overly busy sites confusing. Too many flashy, whizzy things going on might be fine on commercial sites but can be distracting and off-putting in a learning environment. As Gillani and Relan (1997: 236) emphasize, multimedia 'should convey information rather than be an art piece'. Therefore, while interactive multimedia can

> 'Reliance on exciting visuals may distort the curriculum by focusing students' attention on the entertaining and provocative features of the presentation rather than encouraging thoughtful analysis of their underlying meaning.'
>
> SHERRY 1996: 341

enhance your site, and the learning experience of the students, beware of the trap that this can become (Hedberg *et al.* 1997).

Planning construction

Once you embark on your own WBI pages you will be surprised how quickly they accumulate and the whole project grows. For example, the *Social Geography and Nottingham* project now consists of over 219 pages and around 300 graphic elements, including 'clickable' maps, photographs and aids to navigation and orientation; the *e:net* WBI is made up of 37 pages and 90 images; and, even at the relatively early stages of construction, the *Medieval History* project consists of 46 pages and 19 images. Given these kinds of figures, and the amount of information to be handled and organized, it is clear that sound planning of overall construction is paramount. Further, because the material is arranged, and meant to be navigated, in a non-linear fashion, the editorial role is much more complex than in conventional linear arrangements such as books, articles or lecture notes. Inevitably, time will be wasted and wrong paths will be taken, but much frustration, despair, gnashing of teeth and premature ageing can be avoided by taking time at the outset to consider what material is to be used, in what medium it is to be presented, how it is to be linked with the rest of the material, how the user is to progress through the material, what add-ins are to be included and the like. As the site grows, consider adding a site map as an aid to navigation and as means of viewing the overall WBI (Figure 4.2). Also include in your site some guidance as to how the WBI is to be used (Figure 4.3). Here, as well as lots of navigational aids, the user is guided to online libraries, instructed how to evaluate online information effectively and how to cite web sites.

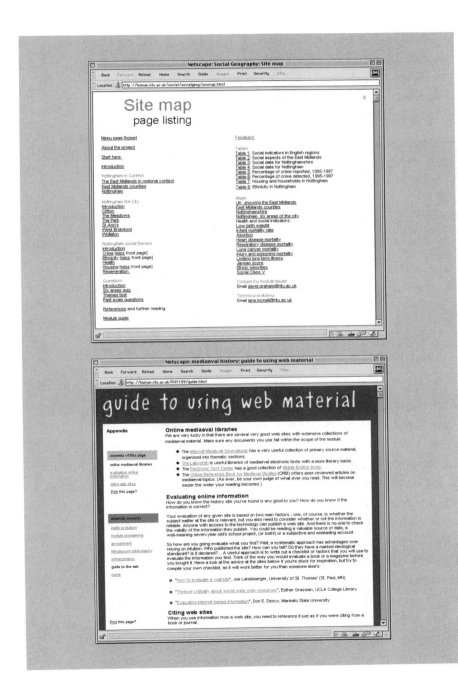

figure 4.2
site map of
the *Social
Geography
and
Nottingham*
WBI

figure 4.3
provide
guidance
on using
the
material

An invaluable tool at this stage is the storyboard (see Figure 2.1). The storyboard not only serves as a basic blueprint but also provides a guide and plan for addition and expansion. Site enlargement is more complex than simply adding pages. These must be added in a logical and technically feasible fashion.

'The damp floor of the Internet sprouted Lecter theories like toadstools and sightings of the doctor rivaled those of Elvis in number. Impostors plagued the chat rooms and in the phosphorescent swamp of the Web's dark side, police photographs of his outrages were bootlegged to collectors of hideous arcana.'

HARRIS 1999: 47

They must be added in some sequence and in a structurally valid way. Links must be sound and viable. Cross-references must be checked. The storyboard provides the skeleton, the steel superstructure of your site. Good storyboarding is analogous to creating robust algorithms. It is possible to rearrange and manipulate the structure at a later date, but a well-thought-out and sound storyboard at the outset can save lots of time, energy and grief in future.

An increasing number of teachers are making use of the internet as a means of delivering course-based material, such as module handbooks, hotlists of useful URLs and the like. Those who have gone on to produce more ambitious, dedicated WBI products like to include links to relevant external sites. We have control over our own material and what we include in our own pages but have no control over what lurks in the web. A lot has been written and said about the quality and educational value of much of what is available on the WWW. Unlike more conventional instructional media, there are no gatekeepers. Despite the dumbing down of the news media in recent years, there is still editorial control which ensures some predictable standards. However, such is the force of the profit motive that students should always be cautioned about the uncritical use of print or broadcast material. Similarly, the exigencies of research assessment exercises in the UK, and similar government-sponsored pressures on researchers elsewhere, have seen a marked decline in the standards of some academic journals. Besides, the gatekeeping procedures in the academy have always been suspect. Comparison with the list of referees and the list of contributors in many journals leads to some interesting conclusions. The 'buddy system' is stronger now than it has ever been. Blind refereeing is very often not that and the minimum publishing unit is the shibboleth of the modern researcher. Funding is so closely tied to publications that findings are falsified and best research practice ignored in the academic community (Roman 1988). The verity and integrity of internet material, then, might not be all that problematic. Nevertheless, it is important to choose your links to the outside world carefully, and it is good practice to check these periodically not only for currency but also for the nature of their content. We are all familiar with the 'HTTP 404 page not found' warning, and not all sites leave forwarding addresses. And there are occasions where the content of a site might change and thereby lose its relevance. There is also the need, during your own surfing sprees, to review newly constructed or newly discovered sites relevant to your own WBI with a view to adding these. Some WBI pages will make more use of integrated links than others. For example, *e:net* makes use of links in a more explicit way (Figure 4.4) than *Social Geography and Nottingham* does (Figure 4.5).

Assessment

Assessment is clearly going to be at the heart of your endeavour. There is a need to build into your pages some means of assessing the fruits of your own labour and that of the students. In many ways assessment and evaluation overlap. The minutiae of evaluation are dealt with in Chapter 9. To avoid undue overlap, we will discuss assessment here only briefly, and in terms of its relationship with the overall planning process.

Untangled web: developing teaching on the internet

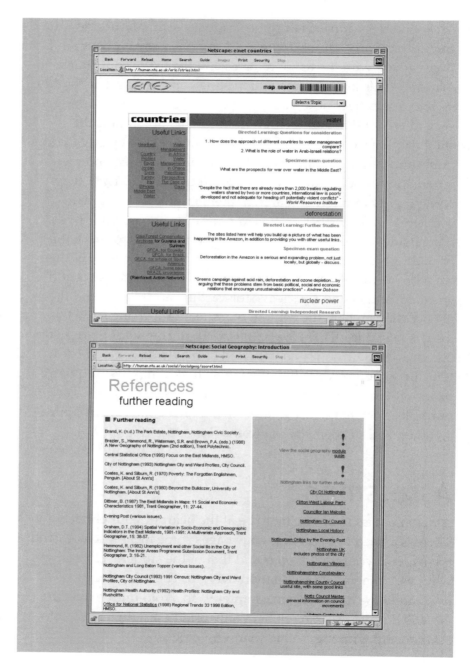

figure 4.4
e:net
makes
explicit
use of links

figure 4.5
links are
used
sparingly
in *Social
Geography
and
Nottingham*

'Technology', as Yong and Wang (1996: 3) note, 'is only a tool and cannot solve poor teaching.' You should not assume that by using the sexiest, cutting edge technology available, students can be fooled. They want message and medium, but are usually sharp enough to distinguish between the two. You must, therefore, assess your own product in a rigorous manner. Fortunately, the technology itself can aid efforts to assessment. From the start, build into your WBI mechanisms for assessment. You will want to include a number of evaluative methods. Have colleagues test-run pages and navigation. Incorporate this personal feedback into improving the structure of the site as well as individual pages. Design into the WBI tailor-made feedback forms that will allow users to provide critical comment on aspects of the site. This should, hopefully, encourage improvement and point out errors of structure and ways in which navigation could be made more user friendly. You can invite users to report back errors of content, spelling and the more detailed aspects of the site. However, as with any anonymous feedback, do not be surprised, disappointed, disillusioned or dispirited if feedback from some individuals is negatively critical, venal, banal or personal. Ideally, feedback from students *should* be anonymous. Participation rates tend to be higher and (most) responses constructive with anonymity guaranteed. There are technical ways of identifying individuals, much as we can recognize individuals in hand-written, analogue evaluation forms. However, avoid this as it serves little purpose, breaches trust and might infringe the Data Protection Act. Credibility with the student body and faculty would suffer and one of the most appealing features of WBI from a student perspective is anonymity (Kapur and Stillman 1997).

Be sure to consider the inclusion of ways in which the learning experiences of the students can be gauged. If the WBI is to be included in the formal assessment of the students then a great deal of thought and planning will have to go into this. Aspects of security and comparability might have to be considered (Hudspeth 1997). For example, how to determine that the student submitting the work electronically actually did it—rather 'than a knowledgeable surrogate recruited by the student' (Falk 1997: 13). You would almost certainly want to discuss these and other issues with a senior colleague or someone in Registry to ensure that regulations are complied with. If the WBI is part of a more traditional course or module, a greater degree of flexibility should be available. Consider the use of tests and quizzes based on the material so that students can test themselves. You may wish to see the results of these, or not, depending on circumstances. We have found that students are very fond of self-assessed tests which involve gadgetry such as radio buttons, and popup and drop-down menus.

Finally, consider ways in which you might get your own feedback to students, individually or collectively. There are a variety of ways of doing this. You might simply report directly to the class in a formal meeting as part of the module or course. This is possible where the WBI is integrated with more traditional delivery systems. You might consider setting up some type of email mentoring, where queries or concerns of individual students can be addressed and relayed to that individual or the whole class, depending on the level of sensitivity of the material. There is also the possibility of making use of bulletin boards as a means of providing feedback and information to the whole group. Also, consider the use of notice-board conferencing. But distinguish between synchronous and asynchronous and the amount of time you feel able to devote to what can be interesting, stimulating, exciting yet time-consuming activities.

chapter five

inspiration, implementation and motivation

'That's the reason they're called lessons', the Gryphon remarked: 'because they lessen from day to day.'

Lewis Carroll *Alice's Adventures in Wonderland* (1865)

Once you have designed and built your WBI package the old maxim involving horses and water becomes an issue. The amount of time and effort that has gone into collecting, manipulating and organizing your material, and then translating it into a well-designed set of pages, means that you will want your students to get the maximum benefit from the learning experience. You will have dealt with inspiration already, that is, your own inspiration during the design and development stages. But it is crucial from the beginning to bear in mind the need to inspire the students. Inspiration is the shibboleth of modern pedagogy. New recruits to teaching at all levels are asked how they would inspire students. Students complain that they are not inspired enough — they do not have enough to do, but when they get things to do are reluctant to do them, unless some type of assessment is involved. The government is threatening to sack teachers who fail to inspire, and so on, and so on. Do not be fooled into thinking that simply placing some material on the web will provide instant inspiration and gratification to the class, either. If anything, WBI requires more planning, with regard to inspiring students, than more conventional learning and teaching methods. In many ways, your own inspiration which goes into the content, media, design, navigation and interactivity of the WBI package will reflect back on the student.

Closely linked with inspiration is implementation. From the outset it is important that the teacher has a clear idea how the web-based material will fit into the overall structure of the course, module or whatever. WBI works best if it is seen to be an integral part of a whole and not just some ill-prepared, last-minute add-on. It is important, therefore, to consider motivation. In this 'sound-bite', 'I want it now', 'what's in it for me', 'instant gratification' society, motivation is everything. Higher education does nothing to discourage this with its promotion of modular, pick 'n' mix education based on continuous assessment. Increasingly, students are only motivated to do something if it is assessed, and some points, credits, certificate or degree can be acquired thereby. Marketability and minimalism appear to have replaced the enquiring mind and learning for its own sake in education and the wider society. Some way of motivating (strongly encouraging) the students to make effective use of your WBI material is therefore a major consideration.

Inspiration

At a time when academics are asked to wear more and more hats, and to keep these on for longer periods, without either the extra money to buy these hats or without their brains becoming overheated, it takes a considerable amount of energy and faith to commit to more time and effort in what might be a largely experimental endeavour. To embark on WBI while doing a zillion other things requires inspiration and enthusiasm. There is also a need to transpose that inspiration and enthusiasm onto a student body which is becoming at the same time more demanding and disinterested. That is becoming an increasingly difficult task.

Yours

To some extent the inspiration involved in the content and design of the WBI will be a reflection of your own interest in your subject area, of your enthusiasm for the web as a teaching and learning vehicle and your own imagination and energies.

If you are lucky you will find supportive colleagues

> 'Web-based instruction is bound to upset the existing system; it will either force change or cause resistance. By bringing new capabilities into existing instruction, however, the Web is redefining the rules and expanding the frontiers of curriculum and instruction.'
> SHERRY AND WILSON 1997: 72

and willing allies within your department, faculty or wider institution whose own ideas, practices and experiences of WBI may rub off on you in a positive way, or against whom you may be able to bounce ideas in a fruitful manner. Using the web for open and distance learning is a long learning curve and can be a lonely experience. As Ptaszynski (1997b) made clear, there is a natural reluctance among staff, in this research-driven era, to invest considerable amounts of time and effort in new teaching methods when these are rarely rewarded in terms of promotion. Not a lot has changed since he wrote those words. Even in the physical sciences, there is a reluctance to invest in WBI. In one UK university, responses towards a pharmacy WBI elicited responses such as 'who is going to do the work for little reward?, 'it would be a huge effort for unknown gain', 'good idea if the time involved in introducing it is recognized' (Sosabowski *et al.* 1998: 28). Some are less pessimistic and foresee a situation where staff will have to become fully involved in new technology (Cornell 1999; Duchastel 1997; Morrison 1997b, 1998). This is the view from the USA. In the UK, while there is a recognized need for '*evolving* lecturers', prepared to embrace the new technologies, such as the web, as delivery systems (Orsini-Jones and Davidson 1999: 38), academic staff continue to be appointed who cannot use a basic word processing package or who order the highest specification and most expensive hardware so that it can gather dust—they must *have* it but do not (or cannot) *use* it. Besides, as Herrmann *et al.* (1999) point out, the 'changing perceptions of the "academic" role and the increasingly pressured environment of the university mean that some lecturers see their interests as lying elsewhere and place a relatively low priority on teaching responsibilities: offers to assist in multi-skilling may be rejected'. There can even be negative aspects to being innovative. Sometimes, according to Rowntree (1998a: 51), teachers 'are "punished" for making their teaching more efficient'. Any time they save in developing WBI could be clawed back by an unsympathetic management that fails to realize the time needed to maintain and develop such a resource.

> 'what young faculty member is foolish enough to spend their most scarce resource (i.e., their time) on an area that will provide no positive return on their investment? That is, help them get tenure. In fact, it may be seen as detrimental as senior faculty do not value "playing with computers".'
>
> PTASZYNSKI 1997B

Do not be surprised or disappointed, therefore, if colleagues do not appear enthusiastic or supportive. There are many reasons why your fellows may be negative, obfuscatory or even downright hostile to your ideas for improving the teaching and learning experience of both students and staff. These include:

- Jealousy—this is bound to occur when new methods are used.
- Fear—not only of the unknown but also of failure.
- Frustration—at being excluded or having to invest time and energy when institutional change for the sake of it makes medium-term plans problematic.
- Distrust—of innovative methods.
- Ignorance—'there's nothing like that on my web'.
- Inherent conservatism—'chalk and talk has stood me in good stead these last 40 years'.
- Institutionalism—lack of support and training.
- Sheer bloody-mindedness.

If our experiences are anything to go by, you will find a mixture of support and suspicion, envy and encouragement, inspiration and indifference. But as long as you remain inspired and positive you will find that the momentum of the project will sustain itself. Once your WBI has taken some shape you will probably experience a 'snowball' effect. The project will expand, you will generate

'Take heart, be positive, and maintain a proactive outlook. The lessons of the future are with us now.... If we do not heed them, we have but ourselves to blame.'

CORNELL 1999: 63

interest among the more dubious of your colleagues and re-energize yourself, once the fruits of your efforts become apparent.

Theirs

One is always hopeful that one's own enthusiasm for something will rub off on others. This wish is amongst the strongest within the teacher–student relationship. There can be few more rewarding experiences within academia when this actually works. However, more often than not, for the class as a whole, this is wishful thinking. As Chapter 9 shows, and as experience tells us, individual students have different experiences of the same learning environments. Some have an intrinsic interest in a subject area or a specific part thereof. Some are particularly keen on the methodology—the extrovert likes presentations, the technology buff loves using hardware and software, the bookworm is happy to plough through the contents of your booklist and so on. There are some students who will get as much out of a course as they can, others do the bare minimum. The skill, the secret, is to inspire as much of the class as possible for as long as possible. Inspiring students to engage fully with WBI is much the same as inspiring them in more traditional modes of learning and teaching. You should never let your enthusiasm for the functionality the web offers get in the way of sound pedagogical practice. The words of Zellner (1997: 351) should always be borne in mind: 'Standard principles of instructional design and educational psychology should be incorporated into any Web-based course design.'

Try to strike a middle ground. Making your pages overly technical and 'difficult' will have a negative effect on the technophobes in the class. Similarly, pitch the level of the material as you would if using more traditional methods. Just because the web offers a cornucopia of presentational methods, do not feel that your pages are lacking or somehow 'not right' because you have not made use of every plug-in and gizmo available. At the same time, make full use of the media available to suit your purposes. This, after all, is one of the main reasons for using the web in the first place. It is the use of multimedia, together with the distance and learning capabilities, that makes web-based teaching so attractive. As well as allowing for more avenues for interesting presentation of material, the capabilities and functionality of the web give flexibility in terms of the way students approach and deal with the information. As Rasmussen *et al.* (1997: 343) have stated: 'Building a repertoire of cognitive strategies to promote knowledge construction for WBI is critical in making learners consumers of information.'

By exploring the various options available in web-based delivery you should be able to engage and interest the students by a measured balance between both message and medium. The web offers the attraction of being able to deliver material when the student wants it (within reason) and in a format that allows the student to work through the material in a variety of ways. The novelty of this should, of itself, enthuse some students. However, as more WBI becomes available, that novelty value will wear off. But it is the interactivity potential of the web that is most likely to inspire the student, if used effectively (Berge 1999; Gillani and Relan 1997). Indeed, it is the interactivity and dynamism of the web that make it such an exciting educational tool (Doherty 1998). By means of interactive tests, quizzes, conferencing, notice boards and the like, a virtual teaching environment can be set up. Many students respond positively to this type of environment (Cummings 1998; Forinash *et al.* 1998; Ptaszynski 1997c). However, because of the presence of 'experts' in some web-based discussion, some students may feel reluctant to participate (Akers 1997).

Like all good two-way teaching experiences, inspiration and enthusiasm should be reciprocal. This is why it is imperative that maximum effort should be expended on building feedback into your WBI, and encouraging users to respond to this. Constructive feedback from an early stage will not only boost your own faith in the project but will help the WBI evolve in a positive and productive manner. In this way the students can, to varying degrees, help in the construction, maintenance and development of the site (Rizza 1998).

Implementation

Getting students to engage fully in something that is not assessed in some way is arguably the biggest challenge in learning and teaching today. The widespread adoption of a modular system based on ever-increasing amounts of continuous assessment does not provide for a particularly enquiring environment. Politicians' and policy makers' obsession with league tables and quantification of everybody's performance but their own helps reinforce the notion that simply having a score is everything and that acquiring something like knowledge for its own sake is a waste of time. How then—now that you have invested all your time and energy in your WBI—do you go about implementing it into your teaching programme?

Rasmussen *et al.* (1997) have proposed what they call the SCo^2T model as a guide to implementation of WBI. This consists of the main elements of Security, Communication, Connectivity and Troubleshooting (Table 5.1).

Table 5.1 Implementing WBI, the main components of the SCo²T model

Level one	Level two	Level three	Level four
Security	Registration	N/a	N/a
	Access	Passwords	N/a
		Inappropriate material	N/a
Communication	Instructors	Feedback	N/a
		Roles	Mentor/facilitator, learner, knowledge constructor
		Assessment	Discussion groups, portfolio, embedded questioning
	Learners	Consumers of information	N/a
		Roles	Teacher, independent, self-directed, knowledge constructor
Connectivity	Access	N/a	N/a
	Systems	Email	N/a
		Distribution lists	N/a
		Support systems	N/a
		Home pages	Multimedia
Troubleshooting	Systems	Email	N/a
		Web pages	N/a
		Distribution systems	N/a
	Support	Orientation	N/a
		Paper materials	N/a
		Helpline	Bulletin boards, voice, email, faq database

Source: After Rasmussen *et al.* (1997) Figure 1

Untangled web: developing teaching on the internet

You will find that you will have to deal with these as you progress, but as we have already pointed out, developing WBI is much more of a circular process than a linear one. As Kessell (1999) stresses, 'the whole notion of a *non-linear* multimedia course is very different from our traditional linear course structures'. This applies to implementation as much as it does to construction. There is no particular ordering of implementation procedures and you may find yourself dealing with password issues in an instructor–learner dimension, as well as from a systems perspective. Troubleshooting and snagging can be linked to connectivity and navigation as much as they relate to communication and content. The succinct guidelines outlined by Barron *et al.* (1996), Montgomery (1998) and Look and Hollar (1999) are still worth observing when designing and implementing WBI.

Security, particularly as it relates to copyright, is dealt with in Chapter 4. The technicalities are discussed in Chapter 8. Communication takes a number of forms. If your WBI is used as part of a free-standing, truly functional ODL package then communications between the tutor and class, between the tutor and individuals and between individual members of the class become a more central consideration than if the WBI is to be used in combination with more conventional methods — as ours are. Thus, it is possible for us to communicate matters of substance, policy or administration directly to the classes concerned, or individual members of the class can communicate more directly with us in a more personal manner than can be effected via telecommunications. That said, we do encourage the use of email mentoring, electronic notice-boards, conferencing, staff–student email and student–student email within our own WBI products. Increasingly, as class sizes rise and as students demand greater accessibility and flexibility, we will all find ourselves making greater use of electronic communication.

Aspects of connectivity are linked to security via access and to troubleshooting via systems. Connectivity is perhaps something most of us would take for granted. In the majority of institutions, there will be others who deal with connectivity issues. Registry will normally deal with the central computing body to ensure that students can access the internet, hardware and other related IT software. In some places this has been devolved to Faculties or Departments. Often, too, basic training in email, web browsers and other IT access issues will be tackled centrally. But do not assume that all students will be able to use the hardware and software. It will be up to you, therefore, to ensure that your students are cognizant of the basics of the hardware and software you will require them to use. First year students might need some guidance, if only in logging on to the network. Some students might be starting late and need support. After first year, there may be students from abroad or other institutions who are unfamiliar with your institution's networking regime. Also, bear in mind the extra support from you and the institution that might be required for students with special needs. Here, you will need to liase closely with personnel in your computing centre—systems administrators, trainers, managers and the like. Be aware of this need, but do not be unduly apprehensive. You will find that there is a great deal of interest in making greater use of the web among 'non-academic' colleagues and, if our experience is anything to go by, you will find several sympathetic and willing people.

Always test your pages on the machines the students are most likely to use. In a truly functional ODL WBI it is practically impossible to check all machines students are likely to use. It would, however, be desirable to give students a list of minimum specifications for hardware, modem speed, software and plug-ins required, as well as any dial-in information, etiquette or usernames or passwords required to effect connectivity. Most likely, students will be using machines provided and maintained by the institution. It is crucial that you check your pages on these. As we pointed out in Chapter 3, you do not always have to design for the lowest denominator machine, but it is always good practice to see what your pages look like on the lowest specification hardware you can find. Check that the required plug-ins are loaded. This is not always unproblematic. Many readers will be familiar with the annoying message, on an ever-increasing number of sites, that plug-in so-and-so is required. You wait patiently as the software downloads. You then install the software to find that not only does it not do what it is supposed to do because it cannot find file xyz.dll, or whatever, but has reconfigured parts of your system unbidden. Like all technology, the web is liable to idiosyncrasy. Download times can vary enormously with seemingly no regard to time or place. Although your pages will probably be resident on a local server, students accessing from outside the institution may find it difficult to access, or download may be interminable, simply because of the esoteric way traffic is routed on the internet. The same applies to any links embedded in your pages. Undergraduates, especially, appear to be more impatient with each successive cohort, and waiting too long for academic-related information to materialize on their screens might not be on their agenda. In this way, the medium is analogous to the international financial system which can reduce several developing countries around the world to financial ruin with a few minutes' trading, yet is unable to cash a standard, personal, international cheque within a month.

It is also advisable that you check that all necessary cards (video, sound and any others) are installed. If sound is available, are headsets provided? If not, students will need to be advised to supply their own, or may wish to do so anyway. It is best to check these things before you start construction. It is pointless having an 'all singing, all dancing' site if you are the only one who can see the dancing and hear the singing. Cross-platform transfer, ever-changing software versions and lack of overall strategic planning of software and hardware developments within higher education institutions not only make the use of the web and other learning technologies problematic, but actually deter individual staff from employing them (Ptaszynski 1997d).

> The best laid schemes o' mice and men Gang aft a-gley
> RABBIE BURNS *To a Mouse* 1786

There will be endless amounts of troubleshooting, especially in the form of snagging pages and dealing with central computing providers within the institution. Snagging is a seemingly endless process. No matter how meticulous your planning, construction and execution, whether rodent or human, some element of your scheme will, as the Caledonian Bard observed, come unstuck. Misspellings and bad links may well come to plague you. We all like our work to be perfect. But at the same time we have all been guilty of the rushed handout or overhead with the embarrassing mistakes. And most of us will have seen the shockingly executed wordprocessed output of colleagues. However, given the open nature of many sites, constructors generally strive for minimum errors with material on the web. That said, mistakes are not always rectified. The internet is littered with poorly constructed and badly maintained sites. Large corporations and other organizations, often employing expensive consultancies, offer examples of sites that have not been snagged completely. One can only assume that the CEOs and other senior managers who commission such sites never get past the home page.

Because of the unpredictability of troubleshooting it is necessary to put in place support mechanisms and some type of feedback. Support can take a number of forms. These include the very simple to the more complex. For example, provide analogue documentation by way of a handout guiding the student through the basics of your WBI, highlighting possible pitfalls and how to avoid these or get out of them. Similar material can be put online, but this presupposes that problems occur once the student has accessed the WBI, or at least been able to access the online help. Do embed as much help as you can within your WBI. Some students will need this, others will need none or very little. Encourage students to phone with specific problems, by all means, but ensure that they do so only during specific allotted time slots. Better to get them to communicate via email for technical, as well as substantive problems or queries. Some of these might be anticipated, so by providing a frequently asked questions (FAQ) database in your pages you will hopefully experience less disruption. It is always very useful and good practice to organize a demonstration of your WBI, preferably with hands-on experience for the students by way of introduction. This will dispel any anxieties some might have and allow for initial teething troubles to be ironed out.

One thing you must be prepared to do when engaged in ODL is to surrender a degree of control. In most circumstances, with traditional methods the lecturer dominates and tries to retain the full attention of the class (Sherry and Wilson 1997). When using WBI for ODL, the lecturer relinquishes much direct control and places the onus on the student to interact meaningfully with the learning materials. Thus, as Roberts (1998: 370) puts it: 'The lecturer's role changes from being didactic to that of facilitator and the student's role changes from being a passive listener to an engaged learner.'

But, as is shown in Chapter 9, and as we would expect, some students will get a great deal out of your efforts, many will get what they can, want or need, and the least enthusiastic, though not necessarily the least able, will muddle through doing the absolute minimum.

Osborne (1995) has identified a number of methods of implementing resource-based learning which apply to WBI. These are:

- Materials are used to prepare for lectures/tutorials.
- Materials are used for practice/revision.
- Materials are used for revision for tests.
- Materials are used as a general resource.

To this we could add:

- Materials are used to introduce and accompany a course or module.
- Materials are used to underpin some aspect/s of the course or module.
- Materials are used to provide a gateway to other material.
- Materials are used as a method of logging student progress.

Our three projects cover most of these. In the *Social Geography and Nottingham* module, for example, the idea of WBI is introduced from the outset and the project is freely available to the students to explore from early on. They are not discouraged from doing this in any way, though it is made clear that time and guidance will be made available later in the module. As a second-level module, three hours of class contact are scheduled a week. Up until week five—around midway through formal teaching—the third hour is used for more formal seminar work. Thereafter, the third hour is free for students to pursue the WBI. Although some students will have looked at the site we try to ensure that they all get some formal introduction to the pages. One of the centrally managed computer resource rooms is booked in class time and two tutors are available to introduce the site and guide students with the material. At this level, only a handful will normally require low-level instructions about hardware, the web, browsers and that type of thing. As the most developed and longest running of the three projects, this is designed to support and underpin many of the topics covered in the module. Although this is integrated into the module as directed learning, it also provides revision, mentoring and gateway functions.

The *Medieval History* project, being geared to a level one, semester one, module requires greater initial support to familiarize the students with the web and the material. As the newest and least developed of the WBI projects, this has been initially conceived as a means of combining the functions of a module guide, supporting information and revision materials. Since the module is team taught, the WBI aims to provide an anchor and cohesive force. Though, at the time of writing, this is largely based on analogue materials it can be developed into a fully integrated, directed learning, multimedia, interactive tool.

By level three, students are familiar, if not all completely comfortable, with the web and in the Environment in International Relations module students are encouraged to explore the WBI on their own. There is, nevertheless, a formal class in the third hour, when *e:net* is first introduced, and tutorial support, thereafter, for those users who might need it. Students are encouraged to make use of *e:net* during class times when a synchronous conferencing session is taking place. This WBI is seen as more of a support to the module, through gateways to further information, online module-related material, conferencing and mentoring facilities.

By encouraging your students to make use of the web in all aspects of their learning, they will develop a familiarity and level of confidence with the medium. And while familiarity might breed contempt, they should be able to make a critical use of web-based material at an early stage in their university education. When you direct them towards your own material, take nothing for granted. Organize classes for them where they all get hands-on experience. For some this might not be necessary, for others it might make all the difference in terms of their use of and reaction to the WBI. But at least one organized and structured introduction to your WBI will help reinforcement — reinforcement that you take it seriously and, therefore, expect them to do the same. Although the main reason for our own WBI packages is that students can access them openly and at a distance, one advantage of leaving aside a time when they should be accessing the material is that scheduling the WBI into a particular timetable slot allows access to a tutor (Sloane 1997). You need not be physically present in the resource room with the students, save for the recommended introductory session/s, but they know that they can reach you in virtual space at your desk, if necessary. The network can be the umbilical cord between you and them.

> 'Though the technology enhances knowledge construction and communication, this is worthless if the student does not understand and use it. Providing students with an introduction to distance learning is of primary importance, and they must be shown more than a "how to" of hardware and software. An introduction needs to include how to learn in the internet/distance environment. This environment is very different for both students and teachers; if that fact is omitted or glossed over, chances of failure increase drastically.'
>
> McVay 1998

Heed the words of McVay (1998). Ensure that the students realize why the information is being presented to them in this way.

Consider making the aims and objectives explicit and putting these on a page (Figure 5.1). Your site should have what Boshier *et al.* (1997: 343) call 'high face validity' (makes sense) and 'credibility'. Enlist the help of colleagues familiar with the site, especially if you anticipate problems with less confident or less able students. Encourage students' continued interest in the site by posting notices there, rather than on the analogue notice-board. Envisage ways in which some kind of directed learning can be embedded within the site. Make references to the material in the WBI in your formal classes, where relevant, but with frequency. Consider ways in which the WBI can be included in some of the formal assessment.

Motivation

As with inspiration, motivation will normally take two different but interconnected forms. One is your own motivation, the other is the motivation of the students. Indeed, motivation and inspiration are very closely linked. What inspires, hopefully motivates. Your inspiration and motivation will hopefully inspire and motivate others. Unfortunately, motivation is particularly problematic in ODL (Cornell and Martin 1997).

As we have mentioned already, there will be points in the development and implementation of the project where you will feel dispirited and disheartened for a variety of reasons:

- Hardware, software or other technical problems.
- You find that some material you wanted to use cannot be used for legal, fiscal or technical reasons.
- Colleagues have let you down in some way.
- You simply have too many other things to do: book deadlines,

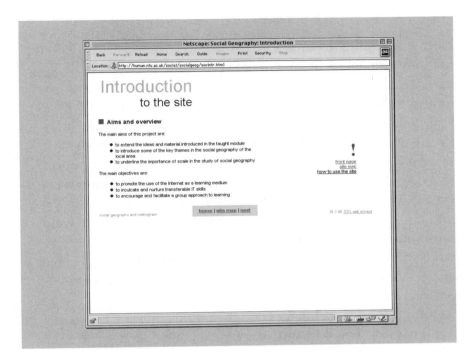

figure 5.1
aims and
objectives
of the
*Social
Geography
and
Nottingham*
WBI

lectures, seminars, tutorials, articles, book proposals, research funding proposals, supervision, meetings, QAA preparation, RAE preparation, staff development, etc.—a life?

- Any one of a thousand other things have gone wrong.

Motivation will wane under these circumstances. This will often be easier to deal with if you are a member of a team—even a team of two will mean that there is support and encouragement when problems arise. Ideas, tasks and expertise can also be shared. Indeed, some of the best models of WBI construction are based on collaboration (Hill 1997; Shotsberger 1997).

'Motivation is actually an umbrella term that encompasses a myriad of terms and concepts (such as interest, curiosity, attribution, level of aspiration, locus of control, etc.); the theories and ideas can be related to individual or environmental and social influences of motivation.'

CORNELL AND MARTIN 1997: 93

Untangled web: developing teaching on the internet

Notwithstanding the occasional problems of morale lapse leading to motivational loss in yourself or your team, once you have invested all this time and effort in your WBI project, how do you motivate the students? Much will depend on the level of the class, your imagination, the nature of the material you are using, the nature of the media you are using and how your pages are integrated into the overall course, module or subject. It is worth thinking about motivation at the outset. Gone are the days when university student motivation was not a major issue, save in a few individual cases. Self-selection ensured that most students were self-motivated — they opted for university and for a course in which they harboured an intrinsic interest for academic or career goals — the lucky few could combine the two. In the current political, economic and academic climate things are more complex. Self-motivation is no longer enough and this must be borne in mind with WBI, as with any other instructional strategy. This is why variety and imagination are integral parts of designing web pages as part of the teaching and learning experience. You must devise 'hooks' (Khan and Vega 1997) to engender interest, retain attention and develop enquiry.

Try to make your pages interesting in terms of aesthetics, as well as content. Aim for some uniformity of design without making the pages boring. Make the home page interesting and informative. The opening page of the *Social Geography and Nottingham* site has a feature (animation) showing several images of the city (City Hall, a mosque, a row of terraced housing, a leafy middle class district) which change every few seconds, as well as a list of contents (Figure 5.2). The *e:net* site opens with a bolder design, but the list of contents is still included (Figure 5.3). Figure 5.4 shows the home page of the *Medieval History* site. This bright attractive page is an attempt to get away from the gloomy, dark designs used in other Medieval sites on the web. You can also introduce subsections of the site with interesting pages. Figure 5.5 shows how a photo-montage provides an interesting lead into the section on ethnicity. The live version is particularly eye-catching and colourful. The other thematic sections—housing, health, crime, urban regeneration—have similar 'front pages'.

> 'Some web courses are an unmitigated bore and represent little more than lecture notes posted on the web. But there is a Hollywood quality to others because, in the desire to be noticed, some authors resort to zany graphics and high-tech glitz and glamour.'
>
> BOSHIER *ET AL*. 1997: 327

There are in practice two aspects to motivation—carrot and stick. We all need a goal or target which is worthwhile to motivate us and pull us in one direction. Very often we need a stick to prod us in the same, or sometimes in another, direction. Sometimes the carrot might disguise itself as a stick and vice versa. But the ideal is to use more carrot than stick.

Carrot:
- Strong links with the module or course. The student must not feel your WBI is just some afterthought tacked on.
- Interactivity. Our students expressed interest in the online quizzes, tests and conferencing facilities we have used. These interactive types of self-assessment, or 'hooks', as Khan and Vega (1997) call them, proved popular.
- Formal assessment. This can be an attraction for some students, especially those more comfortable with cathode ray emissions and keyboard than pen and paper.
- Competitions. Consider introducing some type of competitive element, rather than formal assessment, into your site. For example, the *Social Geography and Nottingham* WBI includes a photograph for which the students are invited to supply an amusing caption. This is returned electronically and a small

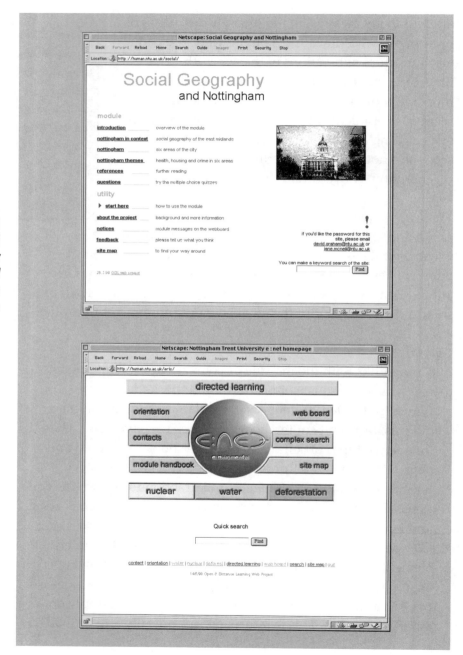

figure 5.2
the *Social Geography and Nottingham* home page

figure 5.3
e:net home page

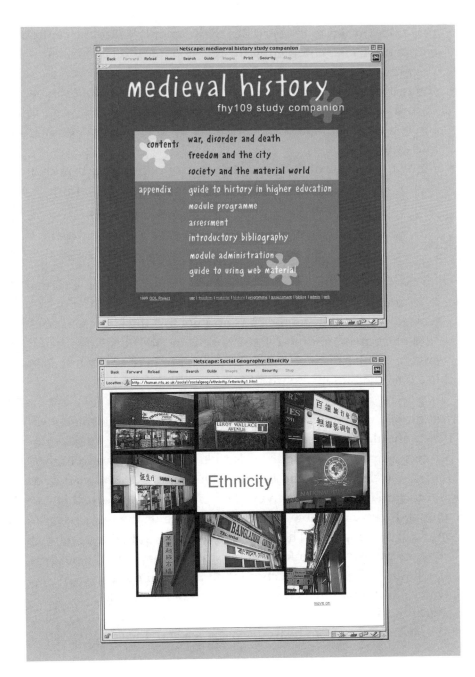

figure 5.4
Medieval History home page

figure 5.5
a photo-montage makes an eye-catching introduction to a topic or theme

prize is awarded to the best. But they can only get to the caption competition *after* completing the revision questions. This has proved very popular.
- Interesting pages. A lively mix of text, graphic and other media encourages users to visit and revisit your site.
- Interesting, current and relevant links to other material, online and otherwise.
- Innovative forms of delivery. Make use of popup or drop-down menus and facts. Use 'clickable' maps and other interesting ways to link information.

Stick:
- Formal assessment can be a carrot, but is more usually regarded by teacher and student alike as a stick. By building some formal assessment element into your WBI you can ensure that it is treated seriously by all the class and not just the interested ones.
- Post all course/module-related notices in the WBI. This should ensure regular visits to the site, though not necessarily intensive use past the bulletin board.

To stimulate motivation, WBI packages, according to Boshier *et al.* (1997), 'should be accessible and involve high levels of interaction (between learners and resources, learners and teachers and each other). As well, they should be attractive.' Figure 5.6 shows how a lot of information can be conveyed on one screen. Here, text is combined with a clickable map to show students the regional context of Nottingham. There is a link to a relevant table of data. There are navigational links to home, the site map and the next page. Figure 5.7 combines a choropleth map of the six areas of Greater Nottingham which form the basis of the detailed geographical dimension of the site. This is combined with a table showing mortality from heart disease in each of the contrasting areas. The shading on the map corresponds to the gradation of the phenomenon under scrutiny. The map and table change as the user selects one of the variables from the row of tabs. The key gives a fuller description of these health and socioeconomic indicators. Note how some of the design features are the same as in Figures 5.1, 5.2 and 5.6. Simplicity and balance, combined with considerable explicit and embedded information, can be effected by means of popup devices (Figure 5.8). Here, a page relating to deforestation from *e:net* provides a box of information alongside popup quotes, popup information and popup facts. Note also the navigational devices and design features.

Table 5.2 lists some of the many facilities/options available to give inspiration to you and your students, as well as making implementation easier and motivation less problematical. These will allow you to spice up your pages, provide interactivity, communicate with the class, allow members of the class to communicate with each other, test the class, allow the students to test themselves and so on. The table lists the ideas you might want to incorporate or explore and the chapter where they are discussed and explained in more detail. Where we have not discussed something in more detail in this book a URL is provided so that you can visit the appropriate site and check it out for yourself to see if your WBI might benefit from the idea.

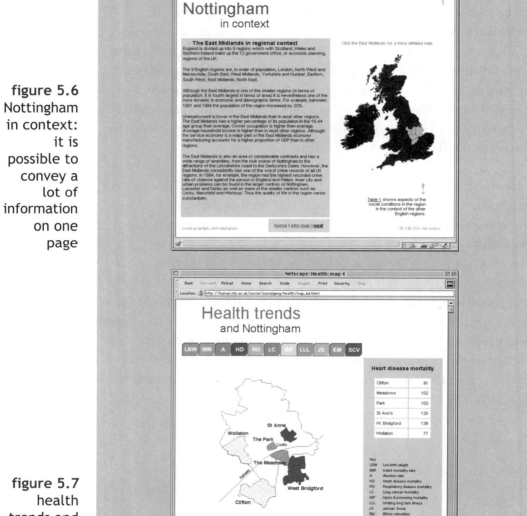

figure 5.6 Nottingham in context: it is possible to convey a lot of information on one page

figure 5.7 health trends and Nottingham: interactivity

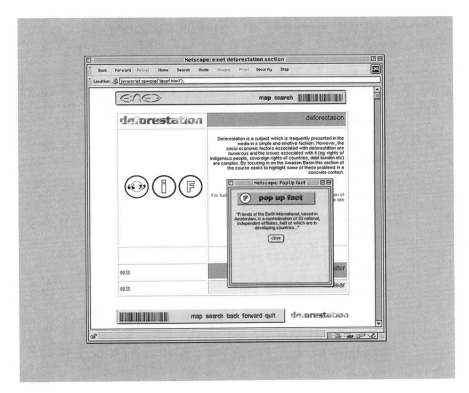

figure 5.8
deforestation:
popup
features
are a
useful
addition to
a page

Table 5.2 Inspiration, implementation and motivation, key to chapters

Idea	Chapter or URL
Animated diagrams	6 and 8
Clickable diagrams	6 and 7
Associative texts	7
Email partnering	8
Email mentoring	8
Online seminars	8
Linked tables	7
Downloadable documents	7
Feedback forms	7 and 8
Popup facts	8
Quizzes	8
Panoramas	6
Virtual fieldtrips	http://www.geog.le.ac.uk/vfc/

chapter six
making graphics for the web

'I warn you, Icarus,' he said, 'you must follow a course
midway between the earth and heaven, in case the
sun should scorch your feathers, if you go too high,
or the water make them heavy, if you are too low.
Fly halfway between the two.'

Ovid, *Metamorphoses*

One of the first difficulties that many people face when building web pages for the first time is how to get graphics for them. This chapter discusses approaches to acquiring pictures for your site, making them into the right format for use on the web, optimizing them and finally, integrating them into your pages. The key to successful web graphics is finding a balance between the quality of images and the speed with which they load into the browser.

Sources of graphics

There are two main formats for graphics in web pages, GIF and JPEG (these are discussed below), but where you find images depends on your budget, how much time you have and, of course, the final effect you need to achieve. You may well decide that you need photographs or drawings to illustrate your material—or graphs, maps and diagrams. You may also wish to produce animation of some description. You will probably also wish to include images in the form of navigation buttons, page dividers, and perhaps typography for main headings.

Online image banks. Type 'free art' into any web search engine and you will be presented with links to a vast number of sites offering images. Many of these are banks of ready-made buttons and backgrounds which are in the right format for instant use in web pages. If you can find something suitable, check the copyright statements before you use them to make sure they are copyright-free. Many will ask you to include a link back to the site from your own as a matter of courtesy. Right-click or click and hold down with the mouse over the graphic you choose and select 'save image' to save the graphic to your hard disk. Other image banks include commercial collections of photographs for which you pay a fee to use.

Computer and design magazines often include small collections of 'sampler' photographs on their cover CDs. These advertise the products of commercial image banks and are often offered free of charge. Again, check the copyright statements to make sure you are allowed to use them online.

Clipart may be of limited application for your site, but some collections include graphics on the theme of education, technology or science which you might be able to use. Clipart will probably not be in the right graphic format to use on the web, so you will need some software with which to convert it. You can buy inexpensive collections of Clipart from high street or mail order computer shops, or sometimes they are supplied bundled with other software. Check the CDs of software you already have to see if any Clipart is included.

Photograph collections on CD vary in price and usefulness. You probably will not want to buy collections or individual images from the upper end of the market, as these are high-resolution images intended for use in promotional material. Less expensive collections of photographs on a mixture of themes are available from high street and mail order computer shops. The photos will often be in the right format for web pages, but you may well want to crop or resize them at least. Some collections also supply software for making limited alterations to the photographs — this may be enough for what you need.

Take your own photographs. This is probably the most appropriate solution if you want to include photos illustrating your subject material and if you have the time. You will need either a conventional camera and a scanner, or a digital camera. Creating your own photographs is discussed in more detail below.

Make your own images. If you have access to drawing or image manipulation software, you can make your own graphics. This is the way to create diagrams or graphs, typography for headers and navigational images. Keep images simple if you are not confident in your artistic abilities — and if you really doubt your drawing skills, imaginative use of typography and dingbats can get you a long way. If you need to produce charts, but do not have access to a drawing package, you could try a spreadsheet instead. Later versions of spreadsheet software like *Microsoft Excel* allow you to save documents as HTML and will export any charts as graphics for your pages.

Ask someone else to do it. If you do not have the time, equipment or software to create your own graphics, you could ask someone else to do them for you. Please be warned, though, that the work may be time-consuming if you need many images. If you are commissioning a freelance artist, it may be costly — and if you ask a friend, his or her patience may wear thin. Prepare as detailed and comprehensive a brief as you can and be ready to compromise a little on your vision. Book a table in a *very* nice restaurant if you've asked a friend to help you and you have a lot of images.

Download time

Many online HTML validation services will also estimate the **download time** of a page, its graphics and other embedded objects. The example in Table 6.1 was calculated by *Bobby*, a validator whose main function is to check a page for its accessibility to people with disabilities. The times are calculated for a 28,800 bps modem. A page loading in five to six seconds at this rate would be very good. The page in Table 6.1 is too slow! What is holding it up is the animated graphic, which alone comes in at just under 40K.

Table 6.1 Download time

Item	Description	Size	Time (secs)
http://human.ntu.ac.uk/social/social.html The page itself		7.77 K	2.16
http://human.ntu.ac.uk/social/pics/socfr.gif A large, animated picture		39.63 K	11.01
http://human.ntu.ac.uk/social/pics/dot_grn.gif A one-pixel graphic		0.05 K	0.01
http://human.ntu.ac.uk/social/pics/Triangle-blue.gif A small, one-colour graphic		0.07 K	0.02
Total		47.51 K	13.20
HTTP Request Delays			2.00
Total + Delays			15.20

Software

If you are sourcing your own graphics, you will probably need to manipulate or adjust many of the images you use before you can use them in your site. You may need to crop them, change their size or colours, or reduce their resolution. The graphics will also have to be in a particular format before they will display in a web browser. So you will need to use some sort of graphics software, unless someone else is creating all the images for your site.

There is a wide selection of graphics software available for most platforms, with a variety of (overlapping) functions and ranging from freeware to the very expensive. Everyone you ask will have their favourite. If you are on a tight budget or no budget at all, with some investigation you may find shareware or freeware that does just what you need.

New software and upgrades of existing graphics software are released all the time, and the capabilities of browsers to display images and animation change with each new version. Before you decide what to use, think carefully about what you actually need and how long you have to learn a new software package. Check reviews in computer or design magazines, online reviews and comments in newsgroups before you purchase expensive software.

Graphic translators and optimizers

If all you need to do is to change an image from one graphic format to another, then use graphic converter or translation software. Many are available as freeware or shareware. Make sure the one you choose incorporates the graphic formats you need. (Formats suitable for the web are discussed below.) You will probably also wish to optimize your graphics for the web, which essentially means reducing the file sizes of the images as much as possible, while retaining as much of the quality of the picture as possible. Graphic converters often offer very useful optimization facilities. *DeBabelizer* is an example of graphics translation software, but there are many others.

Drawing

Vector-based software allows you to draw your own images on screen with shapes and curves. How you use this software really depends on how good at drawing you are. If, like us, your drawing skills are limited, you can used vector-based software for diagrams, simple maps, logos or straightforward buttons and headers. Anything involving gradations of shading and texture may be more difficult to produce. Many people swear by Macromedia *FreeHand* or Adobe *Illustrator*, but other packages are available. Vector-based software also includes 3-D drafting packages.

Image manipulation

Raster-based software maps images onto a pixel grid and is used for manipulating photographs. It can also be used successfully by not-so-accomplished artists to produce complex images. You will need a reasonably fast computer with a lot of memory to make using this type of software bearable and to avoid a long wait each time the image is re-plotted. Adobe *Photoshop* is one of the most widely used examples of raster-based software. Scanners often come with image manipulation software with slightly limited functionality, often 'cut down' or 'light' versions of the major packages.

Web-specific graphics and optimization

Some web-dedicated graphics software offers fewer of the features of the all-encompassing packages, but has added utilities that are useful for producing images destined for the web. These programs may be less useful as general graphics-producing tools, but can save you time in applying effects or textures to images, or alternatively, in optimizing your final pictures. Adobe *ImageStyler* and *ImageReady* and Macromedia *Fireworks* are examples of this type of software.

Graphic utilities

If your budget precludes buying one of the all-encompassing packages, you may be able to achieve the same ends with a collection of smaller, dedicated software. For example, *GIF Transparency* will make one colour in a GIF transparent, so that the background of your web page shows through—if you have an image on a white background, but do not want a big white border round the edge, you can use this software to mask it out.

Freeware and shareware graphics software

Some of the software below we have used, some of it we have not—but if you are on a tight budget, you might find just what you need in freeware or shareware. Some of it is very high quality and indeed rivals commercial software. Remember to virus-check anything you download and stick to well-known sources. Better yet, check reviews and newsgroups to see what other people think of a particular piece of software. This list is just a selection—for a wider choice, you could try one of the download sites like Tucows [http://www.tucows.com/].

clip2gif—Freeware graphic format conversion software for the MacOS.
http://iawww.epfl.ch/Staff/Yves.Piguet/clip2gif-home/

GIF Builder—Freeware GIF animation software for the MacOS.
http://iawww.epfl.ch/Staff/Yves.Piguet/clip2gif-home/
GifBuilder.html

GIF Movie Gear—Shareware GIF animator program for Windows.
http://www.gamani.com/

GIMP—Open Source image manipulation software for Unix-like systems.
http://www.gimp.org/

GraphicConverter—Shareware graphic format conversion software, for MacOS systems.
http://www.lemkesoft.com/

Paint Shop Pro—Shareware Windows graphics editor.
http://www.jasc.com

Transparency—Freeware utility for making transparent GIF files.
MacOS.
Available from All Macintosh.com [http://
www.allmacintosh.com/].

DeBabelizer Lite—Freeware image translation software for the
MacOS.
http://www.equilibrium.com/

Software for RISC OS machines is rather more rare, but Acorn
User's web site is a good place to start.
http://www.acornuser.com/

Graphics animation

There are many solutions to producing animated graphics for a
web site and each has its pros and cons. Most animations and
movies—like those using Macromedia *Flash* or Apple's
QuickTime—require a plug-in to run in the user's browser, and this
may mean that your students have to download one from the
proprietor's site. If they are using an institutional computer, they
probably will not be able to load new plug-ins themselves. Check
what is available before you commit to one technology. If you plan
far enough ahead, you may be able to request that a particular
plug-in is loaded on to your institution's machines.

Animations produce large file sizes and greatly add to the
download time of the page, so it is wise to use them only where
they will be truly useful. Consider what effect you want to
achieve. Do you need a complex animation, with many frames?

Are you animating photographs, or a video clip, or drawings? Do you wish to give the students control over running the animation; that is, will the animation run automatically, or will the students start, stop and rewind it? When you know what you need, read the reviews of software and decide which will best produce what you want. If you are going to include clips from non-digital videos, you will also need access to a computer with a video-in port and some editing software, like Adobe *Premiere*. (Capturing video is discussed in more detail in Chapter 8. Alternatively, if your institution has an audio-visual department, the staff there may be able to offer you help or advice on this.) If you want to produce a simple animation using GIF images (this is not so useful for photographs), you may be as well to draw each frame in one package (Figure 6.1), then animate it using something like the excellent freeware *GIF Builder*. This is just like making a flip-book animation and the format has the added advantage of not needing a plug-in to run. You can also create animation using pre-drawn frames and JavaScript embedded in your web page. (JavaScript is described in Chapter 8, together with more discussion of multimedia.)

figure 6.1
very simple
animation
frames

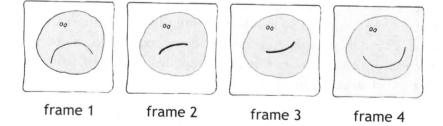

frame 1   frame 2   frame 3   frame 4

Some graphics and multimedia software sites
Adobe http://www.adobe.com
Illustrator, Photoshop, ImageReady, ImageStyler, Premiere

Apple http://www.apple.com
QuickTime, QuickTime VR

Macromedia http://www.macromedia.com
FreeHand, Fireworks, Flash, Director

Doing your own photographs

One of the advantages of web-based delivery is the ease with which full-colour photographs can be incorporated into the material. You may wish to use photographs for added visual interest, to illustrate a concept, to demonstrate a technique or as part of a 'virtual' field trip. You can also use HTML to make a photo act as a link to another page, highlight different areas of a picture as links (an image map), or use a software package to stitch several photographs together to make a 360° panorama.

You will, however, need some photographs to start with. Even the best photo manipulation software cannot make a poor quality photograph into a beautifully clear one, so if you are working with pictures you already have, select only the best to use. However, unlike the production of photographs for print, where the pictures must be of high resolution, creating images for viewing on screen means you can restrict the quality to 72 dpi—the 'average' monitor resolution. You do not therefore need access to the most expensive and highest quality camera or scanner. Pictures taken with a holiday camera will probably provide good enough quality. If you can borrow a tripod, though, you will find it very helpful, particularly if you intend to stitch together several outdoor shots, or are setting up photos of a lab procedure. Try to take outdoor pictures on a sunny day, early in the morning, if at all possible. Indoor pictures will benefit from bright lighting, too, and for close work, photographer's background paper, if you can borrow some, makes for sharper edges round the photographed object.

To transfer standard photographs into digital format, you will need access to a scanner. Scanners of the quality you will need are relatively inexpensive and if you do buy one, choose a flat-bed, not a hand-held. If you do not envisage much use for a scanner in the future, you could always beg time on someone else's — but if you do, plan how you will transfer the pictures back to your own machine, as photographs can take up a lot of disk space. If you want to scan slide transparencies, you will need a specialized scanner. These are available as add-on units for some models of conventional scanner, but can be expensive.

Read the advice in the scanner documentation on how to produce the best results, before you have your negatives processed. The advice may well recommend that you use photos printed on matt, not shiny, paper. When you begin scanning, check the settings are right for display on screen — for example, you will need to scan in RGB colour (not CMYK, which is for photos destined to be printed). Although the pictures will eventually be displayed at 72 dpi, you may elect to scan at a slightly higher resolution if you are planning to manipulate the photos, or if you want to crop out and enlarge a section of the image. You can always resave them at 72 dpi later.

An alternative to taking conventional pictures and scanning is to use a digital camera. Digital cameras store photographs on a data card or disk, rather than on film, so you can download the pictures straight onto your computer. These cameras are becoming better — and cheaper — all the time. You will not need a really expensive one, such as a design agency might buy, but do try to find one with a zoom lens, as these are much more useful than those with a fixed lens. Inexperienced photographers (like us!) will find a camera with a view screen, where the image to be taken is displayed in real time, more useful than one with only a conventional view-finder. The camera will also display the photos taken previously on this screen, so you can selectively delete them if they do not look right. At time of writing, the available digital cameras consume batteries very quickly, so take spares when you go out to take photographs.

Panorama shots

Software like Apple's *QuickTime VR* and like *Spin Panorama* allows you to stitch together a series of photographs to make a 360° or less image. The viewer can then use the mouse to rotate and zoom the image. This has a number of useful applications for learning materials, particularly if you wish to create a virtual field trip or walk-through type scenario. Some packages also enable you to create 'hotspots' in the image that link to other pages or pictures. (The students will need access to a browser with the *QuickTime* plug-in to view the panorama.) If you are taking photographs with a panorama in mind, a tripod is really very useful, even for the most steady-handed photographer. For the best results, take photos that overlap in view by at least 20 per cent. If your camera has a 'wide-angle' mode, do not use it to take the pictures, as this may cause distortions when you load them into the stitching software. Stick to a 35–50mm lens. Figure 6.2 shows how panorama pictures allow the viewer to rotate and zoom the image.

figure 6.2 panorama pictures allow the viewer to rotate and zoom the image

Photo editing

Using photo editing or image manipulation software (like Adobe *Photoshop*) can greatly improve the finished image, whether its origin was a conventional or digital camera. You will be able to crop the image to display only the important part, improve the colour and compensate for poor contrast. You can also, importantly, resize the picture to the size it will eventually appear in the web page. Tutorials for such software can be found online, in computer books, with the software itself and, occasionally, in computer magazines.

Web graphic formats

Common graphic formats like TIFF (tagged-image format file) are not suitable for use on the web, because they make pictures with such large file sizes that they would take a very long time indeed to download. There are several graphic formats that *are* suitable for displaying images in a web page, but not all are supported by all browsers, so most sites use either the GIF or JPEG formats (Figure 6.3). Which of these two you choose depends on the type of image. Producing web graphics is a balancing act between quality and file size — between retaining as much of the picture's quality as possible, and making the file size as small as possible so it will load more quickly in the browser.

GIF

The graphics interchange format (GIF) uses 'lossless' compression to render images with small file sizes, but no loss in quality. If a line in a picture was composed, say, of five green pixels, instead of storing this as 'a green, a green, a green, a green, a green', GIF would store it as 'five greens'. The GIF palette, however, is restricted to 256 colours. What this all means in practice, is that GIF is the format to use for images containing few colour changes, like a line diagram, or a logo, or a picture with flat colours that change sharply. GIF is not suitable for full-colour photographs or pictures with subtly graded colour changes. Mondrian, yes, Turner, no.

JPEG (or JPG)

Joint Photographic Experts Group (JPEG) compression reduces file size by discarding colour information (lossy compression). When you save an image as a JPEG, you can choose how much information to discard, and therefore the resultant quality of the image and its file size. (The choice you are offered will often be on a sliding scale from 'low' through 'medium' to high'.) JPEG can encompass a palette of millions of colours and so is more suitable for photographs and pictures with fine colour gradations.

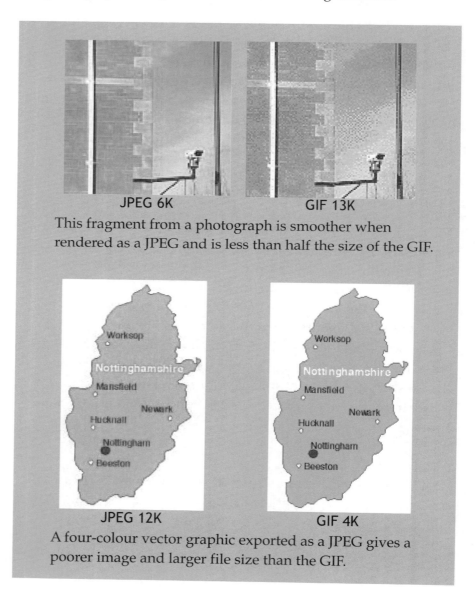

JPEG 6K   GIF 13K

This fragment from a photograph is smoother when rendered as a JPEG and is less than half the size of the GIF.

JPEG 12K   GIF 4K

A four-colour vector graphic exported as a JPEG gives a poorer image and larger file size than the GIF.

figure 6.3
images
rendered
as GIF and
JPEG

PNG

At time of writing, PNG (portable network graphic) format is only supported in later versions of browsers and is not yet widely used, although people who know more about graphics than us argue that it is a superior format to GIF (Veen 1999). Some graphics software will save images in this format, but before you use it, check if your students will have access to a recent generation of browsers.

Optimization

Choosing the right graphics compression format is an important step towards optimizing the look and size of your graphics. There are additional techniques, though, that you can use so that you can incorporate images into your site without causing the pages to load unbearably slowly.

More options for GIF images. If you are using GIF images, you can further reduce their file size by reducing the **number of colours** that make up the graphic, or colour depth. Most graphics packages will give you the option of saving or exporting a GIF in a selection of colour depths, expressed either in the number of colours, or in bits. Some of the later packages will offer a sliding scale, but common pre-sets are 256 colours (8-bit), 16 colours (4-bit), 8 colours (3-bit) and black and white (1-bit). Go as low as you can without compromising the look of the image—this might take some experimentation at first (Figure 6.4). You will also be able to choose the palette of colours which is used to display the GIF—stick to 'adaptive' or 'web' and see which looks better. Reducing the number of colours means that you might get sharply differentiated bands of colour where there were more gradual transitions in the original image. You can avoid this effect by

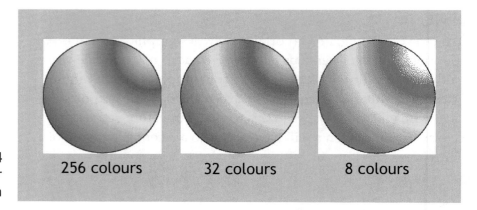

figure 6.4
GIF colour
depth

256 colours   32 colours   8 colours

choosing the **dithering** option when you save the GIF, which will speckle the colour slightly, so the changes are less obvious. Sometimes the image can look worse with dithering switched on, though, so you will have to decide which you prefer.

A further option for GIF images is **interlacing,** which will not reduce the size of the file, but changes the way the graphic loads into the web page. With interlacing switched on, a hazy version of the image will appear quickly as the page is loaded, which will then become more defined while the rest of the page loads. This gives viewers a rough idea of what the picture looks like while they read the rest of the page, but it has been argued that interlaced images take longer overall to load fully than those that are non-interlaced. Also not influencing file size, but very useful for the look of your images, is the option to make a GIF **transparent**. This option allows you to choose one colour in the image that will become transparent, or invisible. The background of your web page will then 'show through' that colour. This is very useful when you are, say, exporting an irregularly shaped object from a drawing package, but you do not want a rectangular border round it.

If the choice of all these options makes you anxious, please take comfort in the prospect that after some initial experimentation, you will quickly find the set-up that you prefer.

More options for JPEG images. There are not so many extra options for JPEG images, as the compression already discards colour according to the quality you select when you save. You may have a choice of making a **progressive** JPEG, which will have a similar effect to GIF interlacing. There is no transparency option for JPEG images, as there is for GIF, so if you are creating irregularly shaped graphics, you will have to set them against a background colour that is the same as your final web page. Figure 6.5 demonstrates masking for different backgrounds.

If transparency is not an option, for a page with a white background, like this one, draw the image against white, so that any extra space around the graphic does not show up. The image on the right would work on a black web page.

figure 6.5
masking for different backgrounds

Stick to the web colour palette. The 'web-safe' palette is comprised of the 216 system colours that Apple and Windows computers have in common. Many graphics packages will include it as one of their set palette options. If you stick to these colours for all your images, there will be a greater consistency in how they display across platforms.

Colours

The 'web safe' colour palette contains 216 different colours. Many graphics and web page programs incorporate this palette, but if you are working in one that does not, guides to the colours are widely available on the web.

The web colour palette is expressed in hexadecimal. It does not particularly need to be expressed so when you are designing graphics, but because all the other elements of a page — background, text, links — must be expressed this way, it makes matching easier if you stick to one system.

Expressed in hexadecimal, black becomes '#000000' and white becomes '#FFFFFF'. To remain within the safe palette, stick to colours with combinations of the following pairs:

FF CC 99 66 33 00

So, for example, a bold emerald green would be '#33CC33' and a rather attractive shade of lilac would be '#CCCCFF'.

Sizing. Make the image the same physical size it will appear in the web page. Photographs can be resized using image manipulation software, then resaved. It is a good idea to resize the photo before you begin to make any alteration to it, and viewing it at its actual size will help you gauge how small or large it needs to be. You will be defining the size of the image in terms of pixels once you begin to build your page, so working in pixels at this stage will obviate confusion later. If you are drawing graphics in a vector package, draw them to the same size they will finally appear.

Small is good. Wherever possible, use graphics that are physically small and have fewer colours. Consider dividing large images up into smaller sections. If you save each section separately, you can reassemble them in the web page.

Reduce resolution. Check that the resolution of your graphics is no more than 72 dpi. As this is the resolution at which most monitors will display the images anyway, saving your pictures with a higher resolution will just be a waste and will increase the file size to no purpose.

Me and my shadow. The special effects that many graphics packages allow you to add to images can be expensive in terms of file size. If you use them, be aware of the extra page-loading time you create. A popular effect at time of writing is to add a shadow behind an image, so that it appears to be a three-dimensional object floating slightly above the page. These drop-shadows can significantly increase the file size of an otherwise relatively small graphic.

Integration and presentation

A recent survey of internet users in England revealed that just under half found it a great frustration to wait for pages with large or numerous graphics to load (quoted in Business Post 1999: 5) The message here is not to overburden all your pages with images. Many sites choose to have a graphics-intensive or animated front page, followed by plainer pages thereafter. If you do decide to include an animated introduction to your site, build in an option for the user to by-pass it. What is engaging once can be tedious after several viewings — your students will not thank you if they have to endure a long introduction every time they log on to check a reference. Allow them to skip the introduction if they wish.

Chapter 7 explains the mechanics of placing a graphic into a web page, and gives the HTML script for defining images. Here, though, is a miscellany of ideas to consider for the integration of web graphics:

Define sizes. HTML allows you to define certain attributes of a graphic as it appears in the final page. Make full use of the attributes you can apply. They include the width and height of the image; explicitly defining these will allow the page to load faster, as the browser can set aside the space for a picture, then get on with displaying the rest of the page while it is loading. If width and height are not defined, the browser must wait until the whole image is loaded before going on to the rest of the page.

Low bandwidth. You could consider creating a version of the image with lower quality and of a much smaller file size, which you can then make the browser pre-load very quickly while it waits to load the 'finished' version. Few sites use this facility, perhaps because it creates extra work for little gain, but we have seen it used to interesting effect — for example, to create the illusion of a fully fleshed object appearing after a 'wire-frame'.

Accessibility. Remember to use the attribute for graphics that allows them to take a brief text description. It is useful not only for students using a screen reader (described in Chapter 3), but also for those who are using the site with the images 'switched off' in their browser.

Keylining. Another attribute you can choose to apply to images is that of a border or keyline. You can make this border as wide as you wish. You might elect to use this option if you are making a graphic a link to another page, or to give a finished appearance to photographs, or a different look to bullets.

Recycle. Once an image has been loaded into the page, it remains in the browser's cache and does not have to be downloaded again for that session. This means that if you reuse graphics on subsequent pages, they will load straightaway. Recycling elements like headers, bullets, buttons and icons not only speeds up overall loading times, but also creates a sense of consistency throughout the site. We have read arguments for and against using *single-pixel* graphics. The idea behind these is that you can change the size of the graphic in the web page, so that the same single-pixel pink dot can be made to display as a bullet 8×8 pixels, or as a divider 3 pixels high by 500 wide. Reusing the same image should speed up the loading of the page, but on the other hand, recalculating its size should slow loading down. We can only say that we have tried this technique, and it made for very fast pages.

Thumbnails. If you need to include lot of large images in your site — for example, photographs of a case study or a lab technique — you might consider allowing students to elect to view them via a menu or picking list. Such a menu would include tiny versions of all the photographs (or 'thumbnails'), next to a brief label, and an indication of the file size of the full version. Each of these thumbnails would then link to a different page, displaying the relevant full-size picture.

Further reading

Bishop, Kevin (1997) Picture perfect, *Demon Dispatches*, Issue 8, July/August,
http://www.dispatches.demon.net/issue8/45a.html

Webmonkey has a very good collection of articles on web graphics, photographs, scanning and software (including reviews).
http://www.hotwired.com/webmonkey/

Web Site Garage is an HTML validator that will also check a page for speed over different types of connection.
http://websitegarage.netscape.com/

chapter seven

HTML and creating pages

He was silent for a few moments, then he said:
'There is a powerful agent, obedient, rapid, easy, which
conforms to every use, and reigns supreme on board my vessel.
Everything is done by means of it. It lights, warms it, and is the
soul of my mechanical apparatus.'

Jules Verne *20,000 Leagues Under the Sea* (1873)

The core tool for creating web pages is Hypertext Mark-up Language (HTML). It is simply a scripting language that defines the elements of a page and how they appear, and unlike actual programming languages, it is very intuitive to use once you understand its logic. HTML, however, was not intended or designed as a page layout tool — so the creation of the type of layout one would take for granted in, say, a desktop publishing package takes a little ingenuity. Like any computer-based language, different versions of HTML are developed all the time, but the current standard version at any given time is delimited by the World Wide Web Consortium. You can find the full specification at its web site, but bear in mind that all the features of the latest version may not yet be supported by all the web browsers. Standardization, or the lack of it, between different brands of browser is one of the more challenging and contentious areas of web development at the time of writing. XHTML promises to improve accessibility to a wider range of platforms, but is not yet the current recommended standard (again, at time of writing).

The elements of HTML script described here are widely supported amongst current versions of browsers, and we have tried to point out where they are not supported by older versions. We also include some script elements that may not be part of the strict HTML standards, but which are supported by many browsers and are widely used.

Software

Although HTML is the basic engine behind web pages, at the time of writing, web page authors, or soon-to-be web page authors, have several choices of what to use to produce it. Each has its pros and cons.

Conversion extensions to regular software. Existing software, like word processing or desktop publishing packages, often includes some form of option to export or save a document in HTML. Using the layout you have created, they translate the document so that it can be viewed in a browser. You do not have to concern yourself with HTML at all. Although these facilities are improving all the time, their interpretation of your layout can be fairly approximate and if you have created a complicated document, it might look very different indeed when you view it in a browser. You may not be able to use this method to produce the higher functions you may need (like forms) and if there are execution errors in the generated HTML, you will not be able to correct them without some knowledge of the language. Additionally, if you use this method, the temptation is to create pages that are really very like their paper cousins, and do not fully exploit the web environment. However, this method is very useful for starting out, or for quickly translating existing long paper documents like module guides or static handouts.

Dedicated web page conversion software has similar pros and cons to the above method, but will probably produce pages that look closer to your originals. This type of software will often also have additional features, and may allow you to index the pages, make a contents page and include a page-by-page navigation system. An example of such conversion software is *Terry Morse's Myrmidon*.

WYSIWYG web page editors. This is one of the most popular methods of web page creation. Editors work on the same premise for web pages as desktop publishing and word processing software do for paper—they help you lay out a page as it will finally look, without reference to what makes the page look like that. So, for example, you do not have to know how text is made to appear as **bold**, you just have to highlight it and click the 'bold' button. You are therefore free to put all your energy into creating the design of your pages, without having to figure out which HTML script element to use. Examples of popular editors at time of writing include Adobe *GoLive*, Macromedia *Dreamweaver* and Microsoft *FrontPage*. A useful basic editor is provided with Netscape *Communicator* — it may well be all you need. Editors often also include options for a selection of JavaScript features, like mouse-activated, animated buttons. Most also let you view the HTML behind the page and alter it 'by hand' if you need to. One of the drawbacks of editors is that very often the finished page does not look exactly as you expected when you view it in the browser (WYSI-**N**-WYG), which can be frustrating. This aspect is improving with every version released, however, as is the range of extra features and higher functions they incorporate — such as directorial control of animations, or site management facilities. Another difficulty is that editors may create the page using a lot of extraneous HTML script, which slows down the loading of the final page. You will also find that they may have difficulty dealing with scripts for additional features you need, and may 'mangle' these if you add them by hand. In fact, many editors have the rather irritating feature of altering any HTML changes you make 'by hand', without you realizing. Finally, with standardization a current difficulty, it is worth noting that some page editors may allow you to incorporate into your page features that do not work on every server or in every browser.

Text editors. Another popular method of page creation, text editors usually give you some help with HTML script, but depend on you having a thorough, or growing, knowledge of the language. You work with the HTML that drives the page, then preview it in a browser to see how the page is shaping up. Text editors vary in the amount of help they give you—some page authors use no-help plain text software, others use packages like *BBEdit* that include pre-sets for most HTML script elements, a web-safe palette and a debugging facility for your script. The great advantage of text editors is that you have complete control over how your page looks and acts, and can include scripts for higher functions without fear of them being corrupted. Another advantage is that it is possible to find very good freeware and shareware text-editing software. Their disadvantage is that you will have to learn HTML—but this may not be a great drawback, as you may find it as quick to get to grips with as one of the larger WYSIWYG page editing packages.

Someone else. As with graphics, you could ask someone else to make your pages. Again, it is a time-consuming process, so consider the cost and prepare as detailed a brief for the site as possible. Remember that with this approach, you will have to commission someone again when you need to add material later.

Which approach you take to web page creation depends on your available time and resources, as well as your level of technological confidence. If you have no budget at all, but some time, you could use the plain text software available on most computers and begin writing pages in HTML. If you have some funding, but little time, you could elect to use a WYSIWYG editor. Most people who make a living developing web pages seem to prefer to use a WYSIWYG editor to initially create the page layout and insert graphics and animations, then they finish off the page and fine-tune it using a text editor.

Alternatively, with pre-preparation, web pages can also be generated dynamically from templates, using a database. This is a useful method for large sites with constantly changing content, like news and magazine sites. You may realistically only be able to use this approach if your institution has its own web server (or if you have a very good relationship with your ISP). If this is the case, ask your server administrator if there are any existing facilities for creating pages this way. Although this method makes for an easily expandable site, it does take considerable time to set up. You will need access to a database (like *FileMaker Pro* or Microsoft *Access*) and probably some software to mediate between the web server and the database itself (such as Blue World's *Lasso*).

Freeware and shareware HTML tools

Some of the software below we have used, some of it we have not — but if you are on a tight budget, you might find just what you need in freeware or shareware. Some of it is very high quality and indeed rivals commercial software. Remember to virus-check anything you download and stick to well-known sources. Better yet, check reviews and newsgroups to see what other people think of a particular piece of software. This list is just a selection — for a wider choice, you could try one of the download sites like *Tucows* [http://www.tucows.com/].

Color Picker Pro —Shareware picker that matches colours to the closest web-safe equivalent. MacOS.
http://www.rootworks.com/cpp/

Eye Dropper—Freeware colour picker for Windows.
http://www.inetia.com/eyedroppereng.asp

BBEdit Lite—Text-based HTML editor for the MacOS that is freeware in its cut-down incarnation.
http://www.barebones.com/

Arachnophilia—Freeware Windows HTML editor
http://www.arachnoid.com/

Netscape Communicator includes a WYSIWYG page editor.
http://home.netscape.com/

Introduction to HTML

In addition to the full specifications for HTML at the World Wide
Web Consortium site, there are countless tutorials and reference
manuals freely available on the web. Look in the listings of a web
directory like *Yahoo* to find one that appeals to you. The best way
to learn how to script in HTML is by practising, with one of these
tutorials or reference manuals open. You can learn the basics in a
matter of minutes, then add to your knowledge as your confidence
with the language increases. Many people learn new HTML
techniques by seeing a page they like, then using the 'View source'
option of the browser to look at the HTML and see how the page
was made.

A selection of online HTML tutorials and guides
Hypertext markup language home page, *World Wide Web
Consortium*
http://www.w3.org/markup/

HTML teaching tool, *Webmonkey*
http://www.hotwired.com/webmonky.teachingtool/index.htm

Sizzling HTML Jalfrezi
http://www.virgin.net/sizzling.jalfrezi/

HTML tag reference at *Netscape*
http://developer.netscape.com/docs/manuals/HTMLguide/index.htm

Reviews by *Builder.com* of many HTML tutorials by can be found at the *Netscape* site: http://webbuilder.netscape.com/computing/webbuilding/powerbuilder/authoring/html/

HTML operates using a system of **tags** or labels which signify how text or an object will act within the page. Most of these tags are quite intuitive once you start experimenting. For example, to make a word appear emboldened, place 'bold' tags around it:

Make this <B>word</B> bold.

This would appear in the web page as:

Make this **word** bold.

The 'bold' tag, like most HTML tags, must actually exist as a pair of tags — one to switch emboldening on, <B>, and one to switch it off again, </B>. You can put as much as you like between the beginning and end tags, as long as you want it all to appear in bold. Tags are not case-sensitive, but many people use the convention of typing them in upper case. The triangular brackets < > indicate that whatever is contained in them is HTML script — whatever is outside them will appear on your page.

All of the familiar text options are available in HTML: *italic text*, **bold text**, bigger text and smaller text.

This is how that sentence would look in HTML:

\<P>All of the familiar options are available in HTML: \<I>*italic text*\</I>, \<B>**bold text**\</B>, \<BIGGER>bigger text\</BIGGER> and \<SMALLER>smaller text\</SMALLER>.\</P>

In fact, HTML is much more useful than just a tool to change the way text looks—if you are typing an address, for example, you can actually specify an address in HTML:

\<ADDRESS>1, The Street, The Town, Countyshire\</ADDRESS>

would appear as:

1, The Street, The Town, Countyshire

You may think this is superfluous until you consider the benefits to students using screen readers.

Some useful tags to know

This is by no means an exhaustive list of HTML tags and new versions of the language, with new, improved tags, will be developed. Remember that the latest browsers at any given time may not support the latest version of HTML. Check which browsers your students are likely to use before you decide to use a newer feature of the language.

As you work with HTML, you will notice that not all tags appear as a pair, but that some exist on their own, and that some basic tags can have additional **attributes**. These are extra instructions for the action of a tag. For example, the tag that denotes the 'body' or main part of the page can take several attributes controlling the background colour of the page, the colour of text and of links. You can add as many or as few of the relevant attributes as you need. Build up tags that can take multiple attributes like this, with a space between each attribute you add:

```
<BODY BGCOLOR="#FFFFFF" TEXT="#000000" LINK="#000000"
VLINK="#000000">
</BODY>
```

If something does not display as you expected, check the syntax of your HTML carefully. If you are using a good text editor, it may well have a checking option which will highlight any mistakes. One missed bracket or closing inverted commas can change the way the whole page appears and may make some elements disappear altogether.

All tags are spelt in American English.

- **Tags that must appear in every HTML document**

  `<HTML></HTML>`   Place these around every HTML document, the opening tag right at the beginning, the closing tag right at the end of the page. They simply indicate that the document is scripted in HTML.

  `<HEAD></HEAD>`   For information that will not be displayed on the page itself, such as the title that appears on the bar of the browser window. This invisible part of the page is called the 'header'.

  `<BODY></BODY>`   This denotes the visible part of the page. Between this pair will be all the content of your page.

- ## Making a title appear in the top bar of the browser window

  <TITLE>The title of your page</TITLE>

  The text you put between this pair will appear above the menu bar of the browser, and in the *Bookmark* or *Favourites* list. This tag goes in the header.

- ## Setting the background of your page

  This is done by adding attributes to the <BODY> tag. If not defined, the default is grey. These settings can be overridden by the browser, if the user wishes.

  BGCOLOR="#FFFFFF"

  Sets the background colour of the page, with the 6 numbers and/or letters after the # (hash) defining the colour in hexadecimal. (In this case, white.)

  BACKGROUND="picture.gif"

  Uses an image on your disk for the background of the page. The browser will repeat the image to fill the page space (tiling). If you want to achieve the effect of a white page with a coloured bar down the left side, you will need to create an image wide enough to fill the width of the browser window, and very thin, so it repeats over and over down the page.

  This image, used as a background, would give a grey page, with a purple side bar.

- ## Setting the default text and link colours

  This is done by adding attributes to the <BODY> tag. If not defined, the default colour for text is black and the default colour for hyperlinks is blue. These settings can be overridden by the browser, if the user wishes.

TEXT="#000000"	Sets the default text colour for the page.
LINK="#33CC33"	Sets the default hyperlink colour for the page.
ALINK="#000000"	Sets the colour of hyperlinks as they are clicked.
VLINK="#000000"	Sets the colour of used hyperlinks.

- **Adding an invisible page description that search engines will find**

  <META NAME="keywords" CONTENT="list the keywords for your page or site in here">

  This 'meta' tag helps some search engines catalogue your pages correctly. This is a solo tag and goes in the header.

  <META NAME ="description" CONTENT="Type a description of your page here.">

  Helps search engines to catalogue your pages correctly, and will be used by some as the description in the listing of your page. This is a solo tag and goes in the header.

- **Making the page automatically move on to another page**

  <META HTTP-EQUIV="refresh" CONTENT="10; URL=next.html">

  This tag makes the browser automatically go to another page, after a brief delay. The number in "content" is the time delay in seconds—to have no delay at all, set this to zero. The file name in "URL" is the page that will be loaded. This is a solo tag and goes in the header.

- **Formatting text**

`<B></B>`	Bold.
`<I></I>`	Italic.
`<EM></EM>` or `<STRONG></STRONG>`	Adds emphasis.
`<CITE></CITE>`	Marks text out as a citation, which will appear in italics.
`<TT></TT>`	Typewriter-style text.

- **Changing the typeface, spot text colour and size**

`<FONT></FONT>`	With the addition of the attributes below, this tag can format the size, colour and face of the text. This is a non-standard but widely used tag. Not supported by all browsers.
`SIZE="–1"`	Taking 0 as default, makes the enclosed text larger or smaller by degrees (–2 to +5). Note: Text size can look very different on different platforms.
`COLOR="#000000"`	Changes the colour of the enclosed text.
`FACE="Arial, Helvetica, Sans-serif"`	Changes the font face. Note: Using this method, browsers can only display fonts that are loaded on the student's computer. List alternatives in order of preference.

- **Making headers**

`<H1></H1>`	Formats texts as a bold heading and inserts a line break before the next text. `<H1>` is the largest heading size; they go as small as `<H6>`.

- Formatting blocks of text

<P></P>   Starts and ends a new paragraph. You can define the alignment of the text by adding the align attribute: ALIGN="right", ALIGN="left" or ALIGN="center".

               A line break, or a 'soft' return. This is a solo tag.

<MULTICOL> </MULTICOL>   Formats text in newspaper-style vertical columns. Can take attributes of: COLS="2" (number of columns), GUTTER="20" (gap between columns) and WIDTH="500"(overall width of column set, in pixels).

<BLOCKQUOTE> </BLOCKQUOTE>   Signifies a block quotation and indents the enclosed text, from both sides of the page.

<DIV ALIGN="center"> </DIV>   Another tag which can be used to define alignment, <DIV> can be placed around large blocks of text, or non-text objects. It, too, can take values of "left", "right" and "center". This is generally preferred to the <CENTER> </CENTER> tag.

- Different types of lists

Bullet list

<UL>   Looks like this:
<LI>Clifton   - Clifton
<LI>Wollaton   - Wollaton
<LI>West Bridgford   - West Bridgford
</UL>

Definition list

```
<DL>
<DT>Apples
<DD>Fruit that can be
red or green.
<DT>Broccoli
<DD>Vegetable that is
very good for you.
</DL>
```

Looks like this:

Apples
  Fruit that can be red or green.
Broccoli
  Vegetable that is very good for you.

Numbered list

```
<OL>
<LI>Boil some water
<LI>Put the tea in
the pot
<LI>Pour the water
into the pot
</OL>
```

Looks like this:

1. Boil some water
2. Put the tea in the pot
3. Pour the water into the pot

- **Making links to other sites or pages**

`<A HREF="nextpage.html"> </A>`
Placed around some text or a graphic, this tag pair creates a link to another page on your site. The text within the inverted commas is the name of the page to which the link points. For 'local' pages, it is a file name.

`<A HREF="http://www.ntu.ac.uk/"> </A>`
This pair creates a link to a page on another site. Here the full URL is included.

`<A HREF="mailto:email@address.ac.uk"> </A>`
Creates an email link. This may be of limited use to students using institutional computers—the function to email from within the browser may well be disabled on shared machines.

- **Making links within a page, or to a specific part of a page**

  You can use page anchors to enable students to navigate around a page. For example, this is how 'return to top of page' buttons work.

  <A NAME="top"> </A> Marks a point on the page, or anchor, and defines the name of the anchor, in this case, *top*. The tag pair can be placed around text, a graphic, or nothing.

  <A HREF="#top"> </A> Makes a link to the named anchor. This pair must, of course, be placed around a graphic or some text.

  An anchor's name can be combined with another file name, if you wish to link to a named point on another page:
  <A HREF="nextpage.html#index">Index of the site</A>.

- **Placing and defining graphics**

  <IMG SRC="picture.gif">

  Inserts the image called *picture.gif*, at the point in the page where the tag is placed. This is a solo tag.

  There are many attributes for graphics, which you can add in to the tag as you need: <IMG SRC="picture.gif" HEIGHT="45" WIDTH="45" HSPACE="2" ALT="Example of residential in-fill">

  It is advisable always to use the WIDTH and HEIGHT attributes, as it speeds the loading of the page, and also the ALT attribute, which you can use to supply a brief text description of the picture. If certain attributes, like that controlling alignment, are not defined by you, the browser will apply its default setting.

HEIGHT="45"	The height of the picture, in pixels.
WIDTH="45"	The width of the picture. Note: Although you can use these attributes to resize a picture, it is not advisable. Using graphics that are very close to the size they will appear on the finished page speeds up page loading.
ALIGN="right"	Sets the alignment of the picture in relation to the page and other objects around it. This can be horizontal — left, right, centre — or vertical — top, middle, bottom. The default setting is left.
HSPACE="2" VSPACE="2"	Creates space horizontally or vertically around the picture, in pixels. This acts a little like 'text wrapping' options in desktop publishing, making other objects stand off from the picture.
BORDER="2"	Creates a border around a picture. Can be set to "0". The default setting is often for the picture to appear with a border of 1 pixel width when the graphic is acting as a hyperlink.
LOW SRC="pic.gif"	Instructs the browser to load a defined image in place of and before the final picture. This 'pre-image' is a separate graphic of lower quality and smaller file size than the picture proper.
ALT="Example of residential in-fill"	Gives the picture a title, in case the student using your page has a text-only browser, a screen reader, or has image loading turned off.

LONGDESC="infill.html"   Not supported by generation 4 browsers, this attribute allows you to link to another page containing a fuller text description of a picture than the ALT attribute allows. The Center for Applied Special Technology (1999) recommends using this attribute and also placing a text **"D"** beside the picture, which links to the same description (a D-link — see Figure 7.1).

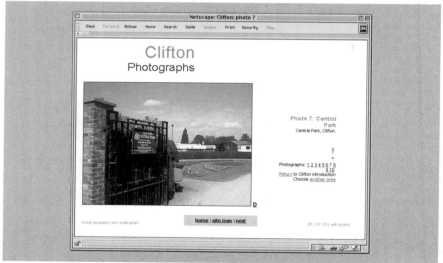

figure 7.1
an image
with a
D-link

- ## A quick and easy page divider
  <HR>

  Inserts a horizontal rule, to match the background colour. This is a solo tag and can take the following attributes:

  SIZE="2"   Sets the height of the rule.

  WIDTH="500"   Sets the width. Can be written as a percentage of the page width (WIDTH="80%"), or as an absolute, in pixels.

  NOSHADE   Creates a plain rule, without a shadow.

- **Making different parts of a graphic link to different pages: image maps**

  An image map is a graphic containing different hotspots that provide links to other pages when the mouse pointer is moved over different areas of the picture. Unlike 'rollovers' created with JavaScript, the different areas of the image map do not change in appearance when the mouse is moved over them. Most WYSIWYG editors incorporate facilities for easily making a graphic into an image map—if you are scripting 'by hand', you will need to define hotspots using pixel co-ordinates.

  `<MAP NAME="name"> </MAP>`
  This tag defines the name of the image map.

  `<AREA SHAPE="rect" HREF="nextpage.html" COORDS="10,20,100,150">`
  Within the `<MAP>` tag pair, you can define shapes which will act as hotspot links to other pages. This example defines a rectangular area. Its location on the picture is defined by two sets of co-ordinates, defining the top left and bottom right respectively.

  `<IMG SRC="picture.gif" USEMAP="#name">`
  The picture you are using as the basis for the map then takes an extra attribute, referring to the name of the map you have defined.

  Put all these elements together and you have something like this:
  `<MAP NAME="nav2">`
  `<AREA SHAPE="rect" COORDS="1,1,71,14" HREF="courses.html">`
  `<AREA SHAPE="rect" COORDS="70,0,143,15" HREF="depts.html">`

```
<AREA SHAPE="rect" COORDS="142,1,212,14"
HREF="resources.html">
</MAP>
<IMG SRC="pics/bar2.gif" WIDTH="284" HEIGHT="14" ALIGN="top"
BORDER="0" ALT="Text links at base of page" ISMAP
USEMAP="#nav2">
```

Working with files and folders

Your web pages will exist as files saved with .html or .htm extensions, finally residing in folders or directories on the hard disk of the server. The images for your pages will exist as separate .gif and .jpeg (or .jpg) files. When you are using <A HREF> and <IMG SRC> tags that make reference to other files on your disk, you need to use a convention called relative addressing to refer to them. In illustration, here is an example file and folder (or directory) structure, as you might have organized it on your disk:

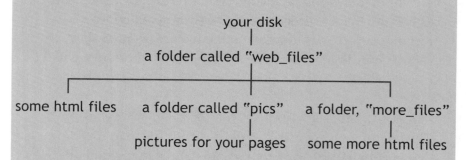

Here's how the *relative addressing* would work:

- A reference to a picture from an HTML file (a web page) in the folder "web_files":
  ```
  <IMG SRC="pics/picture.jpg">
  ```
  i.e. go into the folder called "pics" and use the picture called "picture.jpg".

- A reference to the same picture from an HTML file in folder "more_files":
  `<IMG SRC="../pics/picture.jpg">`
  i.e. go back out of this folder, go into the folder called "pics" and use the picture called "picture.jpg".
- A link from a web page in the folder "web_files" to another page in the folder "more_files":
  `<A HREF="more_files/page2.html">`
  i.e. go into the folder called "more_files" and link to the page called "page2.html".
- A link back from a web page in the folder "more_files" to the home page in the folder "web_files":
  `<A HREF="../home.html">`
  i.e. go back out of this folder and link to the page called "home.html". (If you had to go back two folders, you would simply use "../../", and so on.)

- **Creating tables**
  You can use tables as nature intended, to display tabulated data, or as a page layout tool (Figure 7.2). Be aware, though, that using tables to control the layout of your page can limit the accessibility of your pages to students using screen readers. The HTML for tables is built up from tags defining cells and rows and can be very complex. Take care when making tables to ensure that the number of cells or merged cells on each row matches up, or the table will not display correctly. If you forget to end a row tag, the table may not appear at all.

  Note: if you use a very long table, the loading of the page will be greatly slowed. Try using successive, smaller tables instead.

  `<TABLE></TABLE>`   This is the basic table tag, which marks the beginning and end of the table.

Attributes of the <table> tag

WIDTH="550"   This sets the overall width of the table. Can be defined as a percentage of the page [90%], or as an absolute value, in pixels.

HEIGHT="300"   Sets the height of the table. If you set this attribute to be smaller than the actual table content will fit into, your page may load more slowly. Incorrect table definition can also cause the page to 'flash' while it is loading.

BORDER="1"   Defines the width of the border round the table and its cells, in pixels. If you set this to "0", you will create an 'invisible' table that can be used to control the layout of elements on your page.

CELLSPACING="4"   Sets the amount of space between each of the table's cells.

CELLPADDING="6"   Sets the space between each of the cells and their contents, i.e. how far text and pictures stand off from the sides of the cell.

Table structure: headers, rows, cells

<TH></TH>   Use this if you want a table header cell. The text inside it will appear emboldened and centred.

<TR></TR>   Marks out each row in the table.

<TD></TD>   Marks out each cell within a row.

Cell attributes

ALIGN="left"   Sets the horizontal alignment for text and pictures within the cell(s). Can be set to left, center or right, for a row or an individual cell. The default is left.

VALIGN="top"	Sets the vertical alignment of the contents of the cell(s). Can be set to top, middle or bottom for a row or an individual cell. The default is middle.
NOWRAP	Text lines will not be broken to fit the cell. This attribute does not take a value.
COLOR="#FFFFFF"	Sets the colour of a cell, expressed in hexadecimal. Use this if you want a cell to have a different colour from the background colour of the page. The cell must have some content for the colour to display, so use a non-breaking space (** **) if you want the cell to look empty. This attribute is not supported by older browsers like Netscape 2. If you need to accommodate these, make sure any essential text placed in coloured table cells will also show up against the background colour of the page.

Although usually a table will have an equal number of cells in each row and column, they can be merged if you wish to create different spaces within a table.

COLSPAN="2"	Use with <TD>, if you want a cell to stretch across the space of two or more cells.
ROWSPAN="2"	Use with <TD>, if you want a cell to stretch across the space of two or more rows.

Each of these web pages uses one or more tables as a major element of its layout. The table is most overt in its intended use as a means to display data as in a), but exists behind the layout of the pages in b) and c) as well. Lines of text are prevented from spanning the whole width of the browser window by their placement in a table of fixed size. Table cells control the relative positions of text and graphics, creating a layout grid. Tables are also used here to create blocks of different colour without needing to use graphics.

a) Tabulated data, with coloured cells

b) Using tables to create an invisable layout grid

c) Using tables for low-bandwidth colour

figure 7.2
table
examples

Sample table script

```
<TABLE WIDTH="460" HEIGHT="146" BORDER="0" CELLSPACING="4"
CELLPADDING="4">

<!-- The table header, stretching across all five columns of the table. -->
<TR>
<TH COLSPAN="5">Table 2: Social aspects of the East Midlands</TH>
</TR>

<!-- The first row of the table proper, with headings placed in coloured
cells.-->
<TR>
<TD WIDTH="92" ALIGN="right" HEIGHT="24"><B>County</B></TD>
<TD WIDTH="92" ALIGN="right" BGCOLOR="#33CC33"><B>1</B></TD>
<TD WIDTH="92" ALIGN="right" BGCOLOR="#33CC33"><B>2</B></TD>
<TD WIDTH="92" ALIGN="right" BGCOLOR="#33CC33"><B>3</B></TD>
<TD WIDTH="92" ALIGN="right" BGCOLOR="#33CC33"><B>4</B></TD>
</TR>
<!-- Another row, displaying the first line of data.-->
<TR>
<TD ALIGN="right" HEIGHT="24" BGCOLOR="#0066FF"><B>Notts</B>
</TD>
<TD ALIGN="right">20.5</TD>
<TD ALIGN="right">4.0</TD>
<TD ALIGN="right">68.2</TD>
<TD ALIGN="right">2.39</TD>
</TR>

<!-- A second row of data.-->
<TR>
<TD ALIGN="right" HEIGHT="24" BGCOLOR="#0066FF"><B>Derbys</B>
</TD>
```

```
<TD ALIGN="right">21.3</TD>
<TD ALIGN="right">3.0</TD>
<TD ALIGN="right">71.4</TD>
<TD ALIGN="right">2.39</TD>
</TR>

<!--The final row, with one cell stretching the width of the table. -->
<TR>
<td colspan="5" height="55">
<B>Notes</B><BR>
<B>1</B> % of economically active population in social class I+II
(1995)<BR> <B>2</B> % population in ethnic minority groups
(1991)<BR>
<B>3</B> % housing owner occupied (1991)<BR>
<B>4 </B>average household size (1996)<BR>
 <I>Sources:</I> Office for National Statistics, <I>Regional Trends
33,</I> 1998 and, Central Statistical Office, <I>Focus on the East
Midlands, </I>1997</TD>
</TR>
</TABLE>
```

- **Making separate frames within the browser window**
  Using frames makes the browser subdivide its window into two
  or more sections (Figure 7.3). In each section a different mini-
  page is displayed. Frames are usually used as a navigational
  device, whereby different pages are displayed in the 'main'
  frame, and a constant menu is displayed in a smaller frame —
  often a side bar. They can make site navigation confusing,
  however, and are not supported in older browsers. They can
  also make bookmarking a page difficult, as the URL that will be
  bookmarked will be the initial frameset and not a subsequent
  page the student has viewed within that frame. Search engines

that work by indexing the text on a page can make logistical difficulties for sites with frames. Initially very popular, many people do not now like to use them at all. Make sure the benefits outweigh the potential problems before you decide to use them.

The opening page for a frameset indicates to the browser that frames will be used and which pages or files to load into which frames. The actual content of your pages resides in these secondary files.

<FRAMESET> </FRAMESET> This is the tag pair that indicates frames will be used. The commencing tag goes *before* the <BODY> tag, and the second of the pair goes after the </BODY> tag.

Frameset attributes

ROWS="15%, 85%" This defines how many rows the frameset will have (that is, the number of horizontal frames) and how wide each will be. The width of each is expressed as a percentage of the total browser window width, in pixels.

COLUMNS="20%, 80%" This defines the number of columns, or vertical frames, and their height.

<NOFRAME> </NOFRAME> Placed within the <FRAMESET> pair, this tag will define what will be displayed if the browser your student is using does not support frames. Place it before and after the <BODY> tags and you can include a message or alternative content in the page.

Individual frame attributes

<FRAME>	Each frame is then described. This is a solo tag and can take several attibutes.
SRC="menu.html"	This defines the page that will be displayed in that frame.
NAME="left"	The name of the frame.
SCROLLING="auto"	This attribute controls whether the user will be able to scroll the frame. It can be set to "no", "yes" or "auto". In the latter case, scrollbars would only appear if the content was too big to fit in to the frame.
NORESIZE	Prevents the user from changing the size of the frame.

The resulting frameset script might look like this:

```
<HTML>
<HEAD>
<TITLE>Social Geography and Nottingham</TITLE>
</HEAD>
<FRAMESET COLUMNS="15%, 85%">
<FRAME SRC="menu.html" NAME="left" SCROLLING="auto">
<FRAME SRC="main.html" NAME="right" SCROLLING="auto">
<NOFRAME>
<BODY>
```
Alternate text for browsers not supporting frames. This might be a short message, or the beginning page of a whole 'frame free' alternate site.
```
</BODY>
</NOFRAMES>
</FRAMESET>
</HTML>
```

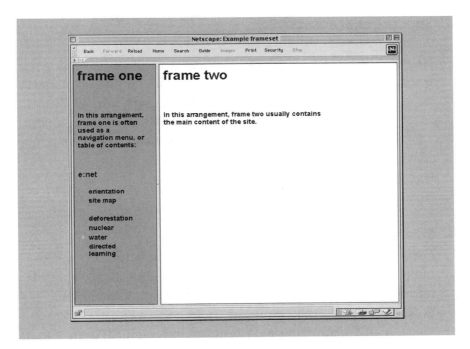

figure 7.3
a frameset

- ## Linking within frames
  The tags for providing links need slight modification if you are
  using frames, so that the browser will load the correct page into
  the correct frame. The <A HREF> tag can take the attribute TARGET:

  <A HREF="nextpage.HTML" TARGET="right"> </A>
  This defines into which of your named frames a linked page
  will load. Define the value as the name of one of your frames, or
  'break out' of frames all together by setting it to
  TARGET="parent".

  You can also make a link open a new browser window by
  setting the TARGET attribute to "new". This can be rather
  disorienting for the user.

- Special characters

&   Ampersand.
                   Non-breaking space. Use it when you need
                        an extra space, as browsers will only
                        recognize one ordinary space (created with
                        the space bar) between two words or
                        objects.
"   Quotation mark.
©   Copyright symbol.

- Forms

Forms allow the page user to enter data into fields, to view and
make selections from lists or drop-down menus and to submit
information back to the server for processing. With HTML,
however, you can only build the 'front end' of a form—the
visual manifestation of fields, buttons and menus (Figure 7.4). If
the user of the page enters data into the form and submits it,
nothing much will happen without a helper application or CGI
located on the server to process the data. These form processing
tools are discussed in Chapter 8; below is listed the HTML
script to insert a form and fields into a page.

```
<FORM> </FORM>
```
Marks the beginning and end of a form.

```
<INPUT TYPE="text" NAME="name" SIZE="30">
```
Creates a one-line text field, with its length defined in pixels. It
is important to give an obvious name to each field, so that you
can remember what each was when you receive data from
students later on. You can limit how much it is possible to type
into the field by adding the attribute MAXLENGTH, as in,

MAXLENGTH="40" (measured in number of characters). If you want some text to appear as a default in the field, add the attribute VALUE, as in, VALUE="Type your email address here". This tag, like many of the other internal elements of the form, is a solo tag.

<TEXTAREA NAME="name" COLS="40" ROWS="5">
Creates a bigger text box, with the width and height defined as columns and rows, in pixels.

<INPUT TYPE="radio" NAME="name" VALUE="value">
This makes a radio button. As they are not much use on their own, use two or more with the same NAME, but different values. This allows for the selection of only one of a set.

<INPUT TYPE="checkbox" NAME="name" VALUE="value">
Used like radio buttons, checkboxes allow the selection of multiple options.

<SELECT MULTIPLE NAME="name" SIZE="5" ></SELECT>
This tag makes a scrolling menu, in which options are listed. The size defines the number of items that will display without needing to scroll. Each option in the list is defined by an <OPTION> tag and the text you want to appear for that item.

<SELECT NAME="name"></SELECT>
This creates a drop-down menu that operates like a scrolling menu, except only one selection can be made.

<INPUT TYPE="submit" VALUE="value">
The button that submits the form to the server, where the CGI will collect it and process it. The VALUE is the text that will appear on the button, usually "send" or "submit".

`<INPUT TYPE="reset" VALUE="value">`
This button resets the form, or clears all the data entries and selections. The VALUE is the text that will appear on the button, usually "clear".

Selected fields from a feedback form

`<FORM>`

`<!--`*The first two fields are single-line text fields. As the* VALUE *attribute has been defined, they will appear with information already typed in them.* `-->`

`<P>`What is your name?`<BR>`
`<INPUT TYPE="text" NAME="FromName" SIZE="40" MAXLENGTH="40"`
`VALUE="Anon"></P>`

`<P>`What is your email address?`<BR>`
`<INPUT TYPE="text" NAME="FromEmail" VALUE="anon@anon.ac.uk"`
`SIZE="40" MAXLENGTH="40"></P>`

`<!--` *In this set of radio buttons, the 'neutral' button would be initally checked.* `-->`

`<P>`The navigation was straightforward to use:`<BR>`
`<INPUT TYPE="radio" NAME="Navigation" VALUE="strongly agree">`
strongly agree`<BR>`
`<INPUT TYPE="radio" NAME="Navigation" VALUE="agree">` agree`<BR>`
`<INPUT TYPE="radio" NAME="Navigation" VALUE="neutral"`
`CHECKED="true">` neutral`<BR>`
`<INPUT TYPE="radio" NAME="Navigation" VALUE="disagree">`
disagree`<BR>`
`<INPUT TYPE="radio" NAME="Navigation" VALUE="strongly disagree">`
strongly disagree`</P>`

```
<!--This is a drop-down menu, with "Please select an option" as the
default selection.-->

<P> Which of the study aids was most useful?<BR>
<SELECT NAME="StudyAids">
<OPTION SELECTED>Please select an option
<OPTION>Self assessed quiz
<OPTION>Scored test
<OPTION>Sample exam questions
<OPTION>Advice in the 'Start here' section
</SELECT></P>


<!--This is a scrollable text box. -->
<P>Feel free to enter any additional comments about the module
in the field below. For example, is there any particular content
you think would be useful? Are there any specific things you liked
or disliked about the project?<BR>
<TEXTAREA NAME="Comments" ROWS="5" COLS="70"></TEXTAREA></P>

<!--These are the submit and reset buttons. -->

<P><INPUT TYPE="submit" VALUE="Send"> <INPUT TYPE="reset"
VALUE="Clear"></P>


</FORM>
```

- Comments

  You will thank yourself six months later if you include
  comments in your HTML script. Comments do not appear in
  the page itself, but do appear in the source HTML. You can use
  them to insert brief descriptions of what does what in your
  page. Put any text between comment markers and it will not

appear in the page:

<!-- *This is a table that holds the main body text.* -->

Comments are also used to 'hide' JavaScript from browsers that do not support it — this is described in Chapter 8.

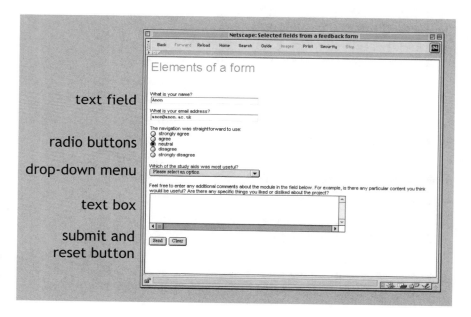

text field

radio buttons

drop-down menu

text box

submit and
reset button

figure 7.4
elements
of a form

- Colour

   The 216-colour web-safe palette is discussed in Chapter 6, where we briefly describe the hexadecimal system used to define colours for web pages. A small selection of colours can be defined using their names — these are red, yellow, green, blue, cyan, magenta, black and white. So to make a page background white, you could use either:

<BODY BGCOLOR="#FFFFFF"> or   <BODY BGCOLOR="white">.
Below is a small selection of hexadecimal values—for a listing
of all 216, you could use colour-picking software, or one of the
many online listings.

#FFCCCC pink
#FF0000 red
#660000 dark reddish brown

#FFCC99 pale orange
#FF6600 orange

#FFFFCC pale yellow
#FFCC00 gold

#99FF99 light green
#00CC00 emerald green

#CCFFFF light blue
#33CCFF mid blue
#000099 navy blue

#FFCCFF pale purple
#CC66CC mid purple
#663366 dark purple

#FFFFFF white
#CCCCCC light grey
#666666 dark grey
#000000 black

Free online HTML validators

Many browsers are fairly forgiving of small script errors and will
still display the page as you expect. Other scripting errors will
completely change the page. **HTML validators** check the script of
your page to make sure that it is correct. Some are more strict than
others!

There are many online HTML validators, some of which will check
more than the HTML scripting alone. Many validators are
commercial services that offer a single page-checking service for
free.

W3C HTML Validation Service
http://validator.w3.org/

Web Site Garage checks your HTML, browser degradability, spelling, speed and graphic size.
http://websitegarage.netscape.com/

Doctor HTML checks a page for HTML, tables and forms, spelling, graphics and links, amongst other things.
http://www2.imagiware.com/RxHTML/

Downloadable documents

You may find that you want to include large text documents in your site and one way of doing this is to include them as downloadable files, rather than as web pages. There are a number of common formats to consider—Portable Document Format (PDF), files created by word processing software, ASCII files and PostScript files. The first two are probably the most easily accessible. If you use word processor files, make sure they are from a software package to which your students will have ready access and ask your server administrator if your web server will recognize this type of document. To create PDF files, you will need *Adobe Acrobat* and your students will need to have access to *Acrobat Reader*. The latter is available to download free from the Adobe web site. PDF files are becoming so widespread that this reader will probably be installed on institutional computers, but check before you commit to this approach.

To link to a downloadable document, use the <A HREF> tag. State the format and size of the file, so the students can gauge how long a wait they will have if they commit to downloading the document:

```
<A HREF="booklist.pdf">Module booklist, pdf, 33k</A>
```

If you are including very large files and want to compress them (zipping or stuffing), make sure that your students will have access to the software to decompress them again.

Cascading Style Sheets (CSS)

Style Sheets are not part of HTML at all, but operate in conjunction with the language and have a syntax all of their own. They work in a similar way to *styles* in word processing and desktop publishing software, in that they allow you to set up rules for the look of a document, which are then applied throughout. By using this tool, you can, for example, define the colour, alignment and typeface of every level of header in your page at the beginning of the document. This leaves you free to use HTML as it was intended — for page structuring, not layout. You can even reference all the pages in your site to a separate document which outlines the style of every element. Then, if you wish to change the look of the whole site, you need only to change this one file. CSS also allow you to lay out the elements of a page without resorting to complex tables.

Using Style Sheets gives you greater control of the layout of your pages, allows uncluttered access for students using screen readers and can save you a lot of time. However, Style Sheets are under-used at time of writing because they are not supported fully, or at all, by earlier browsers. Page authors therefore have the choice of working to ensure their pages degrade gracefully, or of making parallel Style Sheet and non-Style Sheet versions of their sites. If your students have access only to pre-generation 4 browsers, you may find Style Sheets to be of limited use.

CSS tools and guides

Cascade Light — Freeware style sheets editor for the MacOS.
http://interaction.in-progress.com/cascade/

World Wide Web Consortium, CSS Level 1 and CSS Level 2,
specifications and guidance.
http://www.w3.org/style/css/

Webmonkey, CSS Reference and Mulder's Stylesheets Tutorial.
http://www.hotwired.com/webmonkey/stylesheets/reference/
http://www.hotwired.com/webmonkey/stylesheets/tutorials/
tutorial1.html

Webreview.com, Cascading Style Sheets Guide — browser
compatability guides for CSS.
http://webreview.com/pub/guides/style/style.html

Many other articles are available at Webreview.com, including:
Creating Your First Style Sheet, by Eric Meyer.
http://webreview.com/pub/97/10/10/style/index.html

The syntax for CSS is straightforward, but lengthy, and for this reason we do not attempt a full description here, but refer you to one of the many online tutorials. The full specification for Cascading Style Sheets is given at the World Wide Web Consortium site. There are many sites that offer tutorials and guidance in using Style Sheets — look in any of the main web directories or search engines and you will find countless examples, the most useful of which we list here. The CSS syntax is based on three elements; a **selector** (the HTML tag you wish to influence), and **properties** for that selector, each of which has **value**:

selector (in this case, a level 1 header)   *property*: *variable*

H1   (color: green)

Properties for the same selector can be accumulated:

H1   (color: green; font-size: 10pt; text-align: left)

The syntax is related to the HTML of your page in one or more of three ways—inline, embedded or linked. Inline Style Sheet information is applied in the page to one element (like a single paragraph), embedded styles apply to the whole page at the top of which they appear and linked information is contained in a separate .css file, to which multiple pages can refer. This can mean that you have potential conflicts over the style applied to any one selector—in this case, the more 'local' instruction takes precedence. This is the 'cascading' part of the Style Sheets system; inline instructions take precedence over embedded, which in turn take precedence over linked rules.

Further reading

Corcoran, P. (1996) Piecing together server-side image maps, *Webmonkey*, 25 September,
http://www.hotwired.com/webmonkey/

Corcoran, P. (1996) Client-side image maps, *Webmonkey*, 2 October,
http://www.hotwired.com/webmonkey/

DevEdge Online, *Netscape*, http://developer.netscape.com/

The World Wide Web Consortium publishes specifications for HTML and CSS, along with much useful advice, at its web site.
http://www.w3.org/

chapter eight
adding interactivity

He found the tinderbox and took out the candle. He struck
the flint. There was a spark, and in through the door came
the dog with eyes as big as teacups.

Hans Andersen *The Tinderbox* (Haugaard 1976: 15)

In the context of web sites, interactivity provides the user with the opportunity to change or influence an object or environment, or to receive or send feedback. The most fundamental interactivity is provided by hyperlinks, by the process of pointing, clicking and moving to another page. What further elements of interactivity you elect to include in your site is driven both by intent and technological resources. If you have the former, but not much of the latter, you can still make opportunities for meaningful interaction, using relatively simple technologies.

In our experience, observation of and feedback from students has consistently indicated that they enjoy interactive activities in WBI and that they are motivated to use them. This presents an interesting opportunity. Introducing elements of interactivity into your site can create an environment for a greater variety of learning strategies, which can be adapted to suit different learning philosophies and goals. Interactive activities can also provide a useful element of motivation, in the form of practical exercises, reinforcement, revision and reward.

So far so good. The disadvantages of introducing interactive elements revolve around issues of distraction and bandwidth. Having acquired the technology and the knowledge to produce, for example, interactive multimedia, it is very tempting to include it for its own sake. This application can be a distraction from, rather than an enhancement of, learning. Many of the technologies that are used to produce an interactive experience—like JavaScript and Macromedia *Flash*—increase the loading time of pages and can greatly add to the bandwidth of your site.

Before deciding to use interactive elements think carefully about the value they will add. For instance, is a particular item really necessary? Will it achieve your objectives and enhance the learning experience of your students? You should also consider how you are going to implement the particular feature you need. Similar effects can be achieved using various different technologies, and it is advisable to use the simplest method you can find. Many sites employ complex technologies to produce something that could be achieved more elegantly and with lower bandwidth using a simpler method. In other words, do not use a sledgehammer to crack a nut because (with apologies to our feline acquaintance) there is usually more than one way to skin a cat.

How to use this chapter

The main content of this chapter is divided into two sections. The first, 'What do you want to do?', outlines some ideas for interactivity and identifies technologies that can be used to produce them. The second part, 'Selected technologies', briefly describes some of these technologies and gives directions to further resources. Part two can therefore be referenced from part one, thus, we hope, providing you with the shortest route to identifying what you can do and what you need. The information is by no means all-inclusive, as we have only included the more accessible technologies and, of course, new technologies are developing all the time.

What do you want to do?

• Multiple choice question and answer

The simplest way to produce a multiple choice test online is to use **HTML**. For each question, create one page containing the question and the possible answers, making each answer a link to a different response page. If an incorrect answer is selected, the link might lead to a page with some feedback, some more information and a link back to try the question again. You might even include links to the relevant part of your site that discusses the relevant topic. If the correct answer is selected, the page returned could give feedback and a link on to the next question. Thus to complete the test, the student must find the correct answer to each question (see Figure 9.3). Alternatively, you could allow students to skip questions by including extra links. By making 'visited' and 'unvisited' links the same colour, you can stop students seeing the answers other students have selected before them, if they are using institutional computers. It is important to give feedback and further illumination on the response pages to avoid the test becoming an exercise in pointing and clicking. This type of quiz works particularly well where you want to enable the students to test their responses to visual or aural data, or where you want to provide the students with a means of reviewing their knowledge of a procedure or data set. The main disadvantage with this method is that it means you have to produce many web pages for a test of any size. At the end of the test, the students have no quantitative data on how well they did, although for some students this comes as a refreshing change. This type of test has proved very popular in the feedback we have received from students.

• Informal auto-marked quizzes

Alternatively, you may wish to provide the students with a multiple choice test that calculates a mark for their performance. For this, you could use an HTML form and **JavaScript** together to produce a multiple choice test that appears on one page. JavaScript can be used in this context to auto-mark the questions and even to prevent students changing their answers, if you wish. Make sure that the students will have access to browsers that will run

JavaScript before you choose this method, and be aware that including JavaScript in a page will slow downloading.

If you do not wish to become embroiled in JavaScript, you could take a look at the excellent **CASTLE Toolkit** site. The *Toolkit* creates an auto-marked quiz from questions and answers you provide, generating a form-based web page which you copy and paste into your site (see Figure 9.2). The form then returns answers to the server, which calculates a score and returns a page indicating which questions the student answered correctly. Further, you can include short open questions, as long as you define all the possible correct answers. The *Toolkit* program resides on a server at the Leicester University site, but you can apply to put a copy on your own server. Version 2 of the *Toolkit* (in trials at time of writing) will also send the results of each completed test back to a log file on the server, so you can view them later.

CASTLE Toolkit
(Computer Assisted Teaching and Learning)
http://www.le.ac.uk/cc/ltg/castle/
The CASTLE Project is hosted by the Leicester University site and was funded by the Joint Information Systems Committee. The online assessment tools are free for use by Higher Education institutions. The site also offers useful advice (and links) on writing successful multiple-choice questions and more general information on Computer-Assisted Assessment.

- ## Submitted assessment
If you wish the students to submit work to you for informal marking or feedback, there are several approaches to consider.

You can use a **form** and a **CGI** (Common Gateway Interface) script to allow the students to send you answers to open or multiple choice questions. The students answer the questions in form fields on a web page, which is then submitted and emailed back to you. The form can require the student's email address to allow you to post them feedback. Alternatively, the data from the form can be sent to a **database** which stores the submissions and can return a standard response. You could also ask the students to return longer assessments back to you as files attached to an **email** — if they have access to a recent version of a word processing package, they could even save the document as an HTML file before submitting it, allowing you to read the material in a browser and follow any URL references they include.

> 'education at the University mostly worked by the age-old method of putting a lot of young people in the vicinity of a lot of books and hoping that something would pass from one to the other.'
>
> PRATCHETT 1994: 22

These approaches, however, become complicated if you wish to use them for formal assessment, as you cannot rely on their security. A more secure method would require each student to be given a unique log-in identity, which would be returned with their answer — if you use a database, the student could then receive a unique receipt for their work. Even so, obtaining formal assessment in this way could be open to question.

Data sorting exercises

Interactive data sorting exercises can be produced using **JavaScript**. This allows the students to interact with a dataset online — they could, for example, enter values into a table or chart and then see the results or receive automatic feedback.

• Triggered animation

There are many applications for images that change when the

student activates a trigger. The most common application is the navigation button which changes when the mouse pointer is moved over it (a rollover). These rollovers, of course, have a wider use than animated buttons—you could, for example, employ them to change data, or label maps, diagrams and charts. Rollovers require a simple **JavaScript** to work.

Video or animated clips which can be stopped, paused and replayed require software to produce and usually a plug-in or application to play. These 'movies' can be embedded in a web page, or provided as a downloadable file. There is a vast choice of multimedia authoring software, some of which is more widely used for animation, some for video. What you choose depends not only on your platform and on what you want to produce, but on your budget. Read reviews and product specifications carefully before you choose and be aware that many of these programs may take some time to learn. Examples of multimedia authoring and video editing software include *Adobe Premiere* (for video editing), Apple's *QuickTime* (for video editing and sprite animation), *Macromedia Flash* (animation for the web), and *Macromedia Director* (for a variety of media). You can also use *QuickTime VR* or other panorama software to produce a 360° vista from flat photographs, which the student can scroll and zoom. For more on graphics and graphics software, see Chapter 6.

If you are relying on the opportunities available with HTML, you could create a non-animated, triggered *effect* using **image maps** (Figure 8.1). You could use this approach to link areas of a picture to another picture on another page, which then links to another picture, and so on—an example of this might be a series of maps of different scales. Image maps are described in Chapter 7.

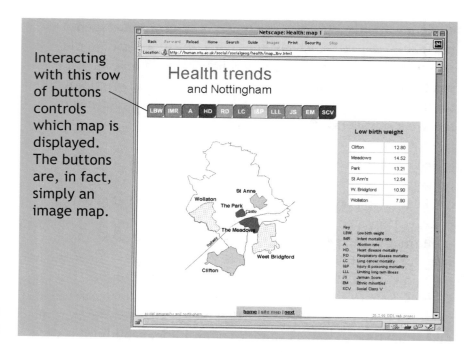

Interacting with this row of buttons controls which map is displayed. The buttons are, in fact, simply an image map.

figure 8.1 image maps

Large graphics files, animations and movies add significant time to the loading of the page, of course, so it is wise to use them only where they significantly enhance the students' understanding or knowledge.

'Despite the hyperbole of its supporters, [the web] will remain an inferior medium for data processing, information management, and other "serious" tasks unless a concerted effort is made to build more powerful interface features into its standard functionality.'

BICKFORD 1999

• Online conferencing, collaboration and mentoring

There are numerous practical applications of online communication between your students, yourself and other people. Collaborative problem solving, co-operative learning, online seminars, guest 'speakers', mentoring systems and role-playing scenarios are just a few. Figure 8.2 shows a typical message board and chat room.

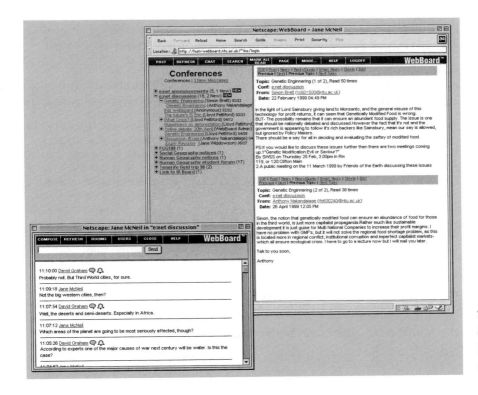

figure 8.2
a message
board and
chat room

Asynchronous conferencing methods include **message boards**, **email**, and **mailing lists**. Using a message board, you and your students can use a browser to post comments and questions to a communal site. Messages are usually arranged into topic 'threads', and you can set up subjects for discussion beforehand. A mailing list also provides a communal environment, where messages sent to the list are circulated to all the list members by email. Email can also be used for smaller-scale interactions, between you and a student, groups or pairs of students, or a mentor and a student.

Synchronous conferencing provides an environment for simultaneous, 'live' typed discussion. These are the **chat rooms** and multi-user dimensions that proliferate online, and which can be used to create a lively learning environment. They work best when a fixed time and date are planned for a discussion and when the topic (with any preparation) is circulated beforehand. You might consider using this approach for an online seminar, or invite a guest 'speaker' for a question and answer session. As the rules of conversational turn-taking are rather different online, it is wise to give some thought as to how the discussions will be structured — or you could encourage the students to develop their own 'rules' (see Chapter 1).

• Popup windows and drop-down menus

If you want to provide optional 'spot' information in separate windows or menus there are two straightforward methods of doing this. The easiest is to put information into a drop-down or scrolling menu as part of a **form**. You do not need any CGIs behind the form in this instance, because the student will not be submitting data — just type the text you want to appear as options in the menu. This is a simple way of including extra information without using up much space on the page. The other approach is to include a link, either from text or a visual clue, that will open up a new browser window. You can do this using a targeted <A HREF> **HTML** tag if you do not need the window to be a particular size. With **JavaScript**, you can make the link open a window of a defined size, in a defined position on screen, with or without tool and scroll bars. You can also give this new window a 'close' button, cause it to close itself after a defined interval or 'remote control' the main window from it. A common application of this type of secondary window is as a site map, which the user can call up at any point and from which they can direct the main window to various parts of the site. An example of drop-down menus is shown in Figure 8.3.

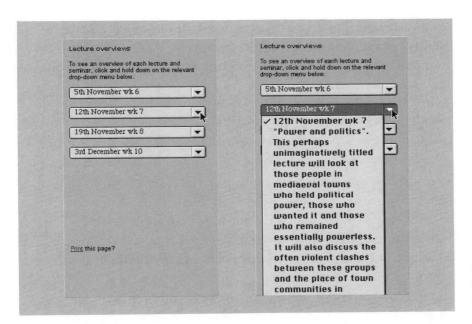

figure 8.3
drop-down
menus

• Text and site searches

Your students can, of course, use their browser to search for key
words and phrases in the page that is currently loaded. If you
would like them to be able to perform keyword searches on the
whole of your site, you can use off-the-peg CGIs which are
available both as freeware online and as commercial software. The
web server you are going to use may already have a search facility,
either in this form, or as part of the web server software, so it is
well worth contacting your server administrator or ISP to ask. If a
search facility on your own server is not an option, you could link
in to remote search facilities available online. Whichever approach
you take, find out how the search engine works and modify your
page structure to give more accurate results, if necessary — some
engines search the whole text of a page, others look for descriptors
in 'keyword' or 'description' <META> tags. These tags are described
in Chapter 7.

Free search scripts If your server or ISP does not have a search facility for you to use, you could try a freeware search script. Remember to check carefully anything you download and stick to well-known sources.
Matt's Script Archive, Inc.
http://www.worldwidemart.com/scripts/
or via *NetFreebies.Net*   http://www.netfreebies.net/

Free search services If you cannot load CGI scripts on to your server, you could consider a remotely hosted search service. These are sometimes provided by commercial companies who offer them as a demonstration or promotion of their services. We have not used one of these services, so cannot comment on their usefulness, but one well-reviewed one is:
Atomz.com — offers a free search facility if your site is under 500 pages. In return you must place their logo on your page and provide a link to their site.
http://www.atomz.com/

• Making interactive frames

Dividing the browser window into separate frames is accomplished with HTML, as described in Chapter 7. The most common use for this is as a menu bar, which controls what pages are loaded into the 'main' frame, but it has wider applications for interactive activities. Four frames could be used to build 'mix and match' windows, which interact with each other — instead of a static 'menu bar' and active 'page', all of these frames could affect what appears in the others.

• Obtaining feedback

If you would like to offer your students the chance to send electronic feedback, you can, of course, invite them to do so by

email. This could take the form of open comments, or of replies to questions you post on the web site itself. Alternatively, you could invite feedback via a **message board**. You may find that you elicit more response if you include a feedback form in your site—this could be a **downloadable document** that the students print out and return conventionally, or an HTML **form** that they submit. You will need a **CGI** to process the form and email back to you, or to send it to a **database**.

Selected technologies
- ## Forms

The syntax for HTML forms is described in Chapter 7, but if you are going to use the form to send data back to a CGI for processing, you must add some attributes to the basic opening <FORM> tag. These are ACTION, NAME and METHOD. The NAME attribute simply defines a name for your form. The ACTION attribute defines what will happen to the data collected by the form—and so usually refers to the CGI script or application on the server that will process it. The METHOD is the way this can be done and can be either 'get' or 'post':

<FORM ACTION="socform.fdml" METHOD="post" NAME="feedback">

If you are using a CGI script provided on the server or by your ISP, or a ready-made one that you have downloaded, check the instructions for what method you should use. (If the students are to use the form to email information—like feedback—to you, 'post' is the method you will usually use.)

You may need to include information in the form that you do not want to appear on the page—such as the address to which the form is to be mailed, or indeed a subject line for the email. This can be done using 'hidden' fields:

<INPUT TYPE="hidden" NAME="To" VALUE="your.email@address.ac.uk">

<INPUT TYPE="hidden" NAME="subject" VALUE="FHY109 module feedback">

- CGIs

CGI (Common Gateway Interface) scripts are, among other things, small applications that handle the data submitted by forms. They reside on the web server and might be written in any programming language that the server understands. If you are going to use forms on your site, you will have to get to grips with CGI, but this is not as frightening for non-programmers as it sounds, as there are many ready-made scripts available.

Before you begin to investigate CGI, though, you need to talk to your server administrator or ISP, because what you need might already exist on your web server. Most good ISPs and many institutional servers will already have popular CGIs set up and ready for you to utilize—form emailers being one of the more popular. Ask what is already available and how you go about using them.

If nothing is already set up, you will need to ask if you are allowed to run your own CGI scripts on the server and how to go about this. Server administrators are understandably cagey about allowing what amount to little programs loose on the server, so be prepared to compromise, or even to find another solution. You have several options if you have to provide your own CGI script:

- Learn to program (not appealing to all, including the authors).
- Commission a programmer to create a script for you (expensive).
- Buy a commercial off-the-peg solution (make sure it will run on your server).
- Use a freeware or shareware CGI (if you can find something suitable).

Several sources of free scripts are listed on these pages — check over anything you download carefully before installing it and running it on the server.

Free online CGI resources

If you are planning to use pre-written CGIs, be careful what you download, stick to well-known sources and check copyright permissions.

National Center for Supercomputing Applications (NCSA), *The Common Gateway Interface* — includes an overview, advice and sample CGI scripts in different languages.
http://hoohoo.ncsa.uiuc.edu/cgi/overview.html

Email.cgi version 5.0 for MacOS — a freeware CGI that will collect submissions from an online form and email them back to you. We have used this and love it!
http://www.lib.ncsu.edu/staff/morgan/email-cgi.html

Matt's Script Archive — a large library of free scripts.
http://www.worldwidemart.com/scripts/

NetFreebies — includes links to free scripts and applications.
http://www.netfreebies.net/

Webmonkey has tutorials on producing forms, CGIs, and working with databases.
http://www.hotwired.com/webmonkey/

- Easy JavaScript

JavaScript is a programming language that provides one of the main ways to introduce interactivity into your site. Unlike Java, it is an interpreted, not compiled, language and so can be incorporated directly into a page. This means that features you include will work 'on the fly' without needing to send information back to a program on the server. It can be used for, among other things, validation and interactivity in forms, navigation, image swapping to create 'rollover' buttons, calculations, displaying the time and date, games, auto-marked quizzes, passwords, clocks, 'select and jump' menus, 'popup' windows, sound, animation, browser detection and plug-in detection.

The main drawbacks of JavaScript are that it slows the loading of a page and is not supported by all browsers. Later versions of the most widely used browsers (Explorer and Netscape) do support it — but they may interpret a given script in different ways and may not support all the features. JavaScript is relatively easy to learn, but it is nevertheless a programming language and will take time to learn thoroughly. This need not stop you from using it, however, as there are countless online tutorials and it is possible to achieve useful effects without having a full understanding of the language. There are also several major sites that provide a vast choice of free ready-made scripts for you to copy and paste into your pages. All these sites usually request in return is a credit or a link back to their site. Several online tutorials and script libraries are listed here, along with Netscape's JavaScript reference manuals. If you learn enough of the language to get by, you can use these resources to source and adapt scripts for your own purposes. In addition, some page editing software, like Macromedia *Dreamweaver*, incorporates JavaScript utilities — so if all you need to do is to make rollover images, you can use their WYSIWYG interface.

If you decide to use ready-made scripts, embedding them in your page is very straightforward:

- The main part of the script is usually placed between the <HEAD></HEAD> tags, after the closing </TITLE> tag.
- The tag to denote use of JavaScript is <SCRIPT>, which operates as a pair and usually includes an attribute describing the language, like this:
  <SCRIPT LANGUAGE="JavaScript">*the script goes in between*</SCRIPT>
- You should 'hide' the script from any browsers that do not support JavaScript, or they will display it as text at the top of your page. You can do this by placing the script inside HTML comment tags:
  <SCRIPT LANGUAGE="JavaScript">
  <!-- *the script goes here, between the 'start' and 'end' comment tags*-->
  </SCRIPT>

- Unless you are quite sure that all your students have access to JavaScript-enabled browsers, do not make vital elements of your site dependent on it (or provide a non-JavaScript alternative).

Free online JavaScript resources

If you are planning to use pre-written scripts, be careful what you download, stick to well-known sources and check copyright permissions.

Netscape's *JavaScript Guide* and *Reference* manual.
http://developer.netscape.com/docs/manuals/js/core/jsguide/index.htm
http://developer.netscape.com/docs/manuals/js/core/jsref/index.htm

JavaScript Source — a vast library of free scripts.
http://javascript.internet.com/

JavaScript *Tip of the Week* archive at WebReference.com — includes a browser compatibility guide.
http://webreference.com/javascript/

Savio, Nadav (1999) How to steal JavaScript, *Webmonkey*, 20 July 1999
http://www.hotwired.com/webmonkey/99/30/index1a.html

Webmonkey has many more resources on JavaScript, including tutorials and a code library.
http://www.hotwired.com/webmonkey/javascript/

- Databases

Databases can be used to receive information from online forms, mark quizzes, return dynamically generated pages and provide searchable catalogues or article libraries. (Sites you may have encountered that incorporate web-based catalogues or online shopping are, in all probability, run this way.) Using a database significantly extends the functionality you can include in your pages. This approach does, however, take a considerable effort to set up the first time you use it. The first step is to decide what exactly you need to achieve, then speak to your server administrator or ISP. There may well already be a database on your web server which you can use. You will have to get to grips with the syntax that interacts with the database (this is usually fairly straightforward) and acquire a good understanding of fields and database operations, if you do not already have one.

If using an existing database on the server is not an option, and it seems unlikely that this service will be available in the near future, you may be able to take a DIY approach, using your own computer as a mini-server. This is only feasible if you connect to the internet through an institutional network (LAN or WAN), not through a dial-up service on a modem, and of course, you must be able to leave your computer switched on 24 hours a day. Using a database package with in-built web features, you can create page templates that will serve pages to the web from your own machine. You can then link to your database-generated pages from your main site using an <A HREF> tag. Instead of citing a URL, however, you insert the Internet Protocol (IP) address of your computer:

```
<A HREF="http://123.45.67.890/"> </A>
```

(If you are not sure what your IP address is, look in the network settings in your computer's control panels.)

We have used this approach to serve databases and dynamic pages to the web from a reasonably speedy desktop computer. We used *FileMaker Pro* as the database, *Simple Text* as the page template editor and chocolate as the development fuel.

- Message boards and chat rooms

A web-based message board is a series of pages where your students can read messages arranged in conversational threads or topics and post their own messages or replies. There are many examples of these on the web and you can either create one yourself, or use one at an external site. The former option is preferable if you want more control over the message board environment. Boards can either be open, where anyone can log in, or closed, where only pre-defined users have access. If you want to set up your own, you can do this with a server and dedicated software (we have used O'Reilly's *WebBoard*), or with an application that generates dynamic pages, like a database. The former is, of course, the more expensive option, but will offer greater functionality.

Shareware and freeware options for message boards and chat rooms

As always, be careful what you download and stick to well-known sources. If you do find that this approach looks the most appealing, note whether the utility or service you are considering requires the students to have access to a particular program.

Look for freeware and shareware utilities and services via:
TUCOWS http://www.tucows.com/
NetFreebies.Net http://www.netfreebies.net/

Rapidly increasing in popularity at time of writing are facilities for live 'voice' conversations. The students will need to have access to the voice software to make use of this technology.

A chat room is an online environment where 'live' text conversations can take place. Again, there are many examples online, some using web-based interfaces, others using Telnet or shareware Internet Relay Chat (IRC) software. Conversations take place in a text window, where users type in what they want to say and when they submit it, it becomes visible to the other people logged on. 'Live' is sometimes a relative term—if the speed of connection is slow, there can be considerable lag between replies, necessitating a new approach to conversational turn-taking. This type of environment ranges from simple 'one room' conversations to complex multi-location landscapes—one-topic discussions or varied Multi-User Dimensions. As with message boards, you can set up your own, or use established services online. Your institution or ISP may already have these services set up, so it is worth investigating what is available.

- ## Multimedia

There are many different formats you can use to include video or audio elements in your pages and each has its advantages and disadvantages. The main considerations for any given format are whether your students have access to the relevant player or plug-in and the size of file you will create. In addition to format, you also have a choice in the way in which you incorporate the file. You can provide a link to a video or audio file as you would to any downloadable file, using the <A HREF> tag, in which case the students must have access to the relevant media player. Alternatively, you can embed the file in the page itself. Using this approach, a video will appear on the page much like a graphic, and a sound file will start to play when the page has downloaded. The browser the student is using must have the right plug-in loaded for this to work.

The tag to embed looks like this:

<EMBED SRC="video.mov">

It can take various attributes, depending on whether it is a video or sound file that you are embedding. For video, include HEIGHT and WIDTH attributes. For both sound and video, you can include other attributes which determine how the file will play. Each takes values of TRUE or FALSE:

AUTOPLAY determines whether the file will play automatically, or whether the user must click a button to activate it.

LOOP directs the browser to play the file over and over.

CONTROLLER allows the user to click a button to stop or pause playing.

If you intend to include multimedia of any description, consider your students with disabilities and provide captions or transcriptions for audio elements, text descriptions for video, or some equivalent representation. The National Center for Accessible Media (NCAM) site has very useful advice, with examples, on how to do this successfully.

Recording sound

You can record sound on your computer with a microphone and sound sampling software. Your computer probably already has this type of software installed. Sound can also be recorded from a CD ROM drive and also from a cassette player, video or TV, if you obtain the right connector cables (SCART leads, or cables with RCA jacks at one end). The cables are easy to obtain—borrow them from your AV department, from your hi-fi obsessed friend, or buy them at a high street electronics store. You can record very good quality sound, but be aware that sound files are very big. Do check on copyright before you record anything from a commercial source.

MIDI (Musical Instrument Digital Interface) technology allows you to compose, play and record music. If a synthesizer keyboard is really not an option, you can use software which includes samples and use the computer keyboard to compose music instead. (We have created music in this way, but as we have absolutely no musical aptitude, the results were rather frightening.) MIDI files are very small indeed, but played back on most computers sound rather cheesy — like the sound tracks to early computer games. File formats you might use to deliver sound include WAV, MIDI and AIFF.

Capturing video

As with sound, video clips can be captured from video players or TV with the right connector cables (above) and software like *QuickTime*. You may choose to edit it using software like *Adobe Premiere*. When you save the final version, use the compression options of whatever software you have access to. Otherwise, the size of the video file is determined by the frame rate or smoothness of playback, measured in frame rate per second (fps). Keep it at 15 fps or below (television works at 30 fps), but not too low or the video will be unbearably jerky. If the software you are using allows you to define 'key frames' (which allow the viewer to skip to different parts of the video), use them sparingly as they slow down playback. Shareware and freeware editors are available, if rare, so it is worth searching the online archives. For example, *MoviePlayer* 2.5 came free with the MacOS and had basic editing features that later free versions did not — you should still be able to find a copy (Pogue and Schorr 1999: 887).

At time of writing, the main multimedia players are *QuickTime*, *RealPlayer* and the *Windows Media Player*. The *Shockwave* plug-in is also widely used. Check which your students have access to before

Untangled web: developing teaching on the internet

you decide which one to use. File formats you might use to deliver video include .mov, .avi and .mpg.

Restricted access

There is no such thing as a secure computer — unless, perhaps, you never plug it in. However, you may find you want to restrict access to some material on your site, perhaps, for example, to meet copyright stipulations for material you have been allowed to use. You cannot rely on any method to guarantee security of information, but you can prevent most casual visitors from accessing the material, if that is what you want. If you are using your own institution's web server, you could ask if it has the facility to restrict access based on passwords. If your students will only use the material from institutional machines, your server administrator might be able to restrict access to an intranet. Alternatively, you could use JavaScript to require students to enter a password to see certain pages. Or you could make downloadable PDF documents with *Adobe Acrobat*, and use that software's password facility. On the other hand, unless you find you absolutely must restrict access to some material, you might prefer to leave it open in the spirit of freedom of information.

Further reading

Bostock, S.J. (1997) Designing web-based instruction for active learning, in B.H. Khan (ed.) *Web-Based Instruction*, Educational Technology Publications, Englewood Cliffs, NJ, pp. 225–230

Gillani, B.B. and Relan, A. (1997) Incorporating interactivity and multimedia into web-based instruction, in B.H. Khan (ed.) *Web-Based Instruction*, Educational Technology Publications, Englewood Cliffs, NJ, pp. 231–237

Pearce, D (1996) Choosing the right video format, *Webmonkey*, 30 Oct
http://www.hotwired.com/webmonkey/html/96/44/index2a.html

chapter nine

testing, adapting and evaluation

Behold I do not give lectures or a little charity,
When I give I give myself

Walt Whitman *Song of Myself* (1855)

W orking through the process described in this book, you will hopefully be delighted to see before you a fully developed set of integrated, interactive, multimedia web pages with which to delight students and enhance the learning and teaching experience. Having thought about your rationale, content and design you will be pleased to have reached the stage of letting the students use the organized materials in creative and interesting ways. Hooray! Unfortunately though, not quite! This chapter looks at a range of issues related to ensuring that your pages provide the tools needed by students and that this in turn leads to solid learning outcomes. Things do not always work as well in practice as in theory and this chapter begins with the perhaps obvious, but tempting to ignore, advice of Barron *et al.* (1996) to always pilot test your pages. Before releasing your web pages to the rest of the world, conduct a pilot test; watch as a sample of the target audience navigates through the pages and make revisions if they are necessary, or use colleagues. It is almost inevitable that you will encounter problems but it is essential to remain positive and proactive (Cornell 1999). This is a very important stage of the process and doing it well can avoid all manner of future frustrations. If we are talking about the task of evaluation, this can be regarded as something done actively within your project, including the important task of keeping up with a dynamic process and somehow organizing, analyzing and learning from it (Ravitz 1997). Evaluation is also a retrospective activity of reflection once the programme has run its course. It is indeed difficult to draw dividing lines between different stages of the evaluative process. However, this chapter is concerned with what happens *before* students have used your pages 'in anger', and largely, of course, to ensure that your students do not literally use your pages in anger.

It should always be remembered that structures need to be put in place to allow you to test, adapt (revise) *and* evaluate your pages as they grow, on an ongoing basis, rather than only at fixed points in time. This is what is meant, then, in suggesting the difficulty of dividing up the evaluative process. Rather than regarding testing, adaptation (revision) and evaluation as linear processes/ constructions it is perhaps better to visualize them as circular and ongoing. To emphasize this point, some authors put evaluation before revision, and in a sense testing is simply a form of evaluation (Sherry 1996).

Unfortunately then, there is actually very likely to be a certain degree of slippage between your logical planning and how the pages actually work in practice. Therefore, suitable formative evaluation techniques are essential (Nichols 1997). If nothing else, students may use your pages in ways you had not envisaged, though this could, of course, have positive as well as negative outcomes. Students may fail to see what you are getting at, or (and this can be a particular problem with synchronous web sessions) their enthusiasm may get in the way of effective teaching and lead to a need for the setting up of agreed user protocols. Others have already done work in terms of overcoming such problems (Leahy 1999). Web-based learning packages are not unusual in this sense. The only way to ensure that a final product is as good as it can be, and used as you would want it, is by careful and methodical testing. To a certain extent this can be done by you, rather in the manner of proofreading a book. Each and every link needs to be checked and the pages used as if with fresh eyes. Especial care needs to be taken with the use of icons; whilst these provide a very useful shorthand to the initiated, they can be very confusing. You may feel frustrated if someone fails to 'get' your simple icon but not as frustrated as the person who fails to see the 'simplicity' and 'if you can think of a hundred things that can go wrong, and factor them into your plan, you will be struck down by the hundred and first.'

Boyd 1998: 365

is therefore unable to navigate around your pages.

Due to the foregoing, there is no substitute for getting a small group of fresh eyes to test your pages for themselves. Physical separation when the package has gone live means that distance education packages must be carefully designed to eliminate problems as much as possible beforehand (Harmon and Hurumi 1996). There are various ways that can be used to acquire qualitative data which can then be used in adapting your web pages. Roberts (1999) suggests class observations, lecturer interviews, student diaries and student questionnaires for instance. This chapter looks at testing and adapting your product prior to its initial use in teaching. It also explores the various evaluative tasks which are a necessary part of WBI construction. These procedures are part of a continual process. Therefore, once a product has been used by one cohort of students it will not only be evaluated, tested and adapted at the end of the course, but throughout the life of the product.

Testing

Unlike more conventional teaching products WBI must be thoroughly tested. If the site is small, then the whole product can be tested. But one useful strategy, particularly for larger WBI packages, is prototyping (Boling and Frick 1997; Frick et al. 1997; Voithofer 1997). Here, the parts of the WBI are tested, or an initial version can be test-run. This can save large volumes of grief at a later date. Whether you prototype or go with the Full Monty, it is crucial that the product be subject to rigorous testing. One way that a WBI creator can gain feedback and assess his or her product is to have colleagues test the site. You will find colleagues (apart from the technoluddites) generally supportive, encouraging and enthusiastic. If our experience is anything to go by you will find that there are colleagues willing to spend time to test and appraise your WBI in a critical, yet positive, manner. Postgraduates also make useful guinea-pigs. However, as Ravitz (1997) suggests, the items under review in any testing procedure should ideally be done by people who have a stake in the project or else quantity may be emphasized over quality in any review. Moreover, the tester should be particularly wary of the meaning and interpretation people ascribe to the web-based learning product. Those who feel positive may be more willing to contribute to a testing phase resulting in the omission of a perspective from those with more negative predispositions. A further point is that with WBI packages which may literally invite exploration, testing will not always be able to ascertain what actual students are likely to learn. With these caveats in mind, testing is still a crucial stage in the development of ODL pages for the web. It cannot ensure that everything is corrected but it can help to ensure that pages function efficiently if not perfectly. The testing procedures should address the following, at least:

- That navigation is logical to as many users as possible (not as easy a task as it might first seem).
- That structural integrity is maintained, that as many bugs and snags as possible have been fixed prior to general use. You cannot rely on class members to inform you of these. Some may be helpful and constructive—many will fail to notice any but the most major of snags.
- That internal and external links are stable and live.
- That internal loops and culs-de-sac have not been created.
- That all interactive elements are that.
- That your pages work effectively on the most likely common platform.

Clearly, testing early in the process is important so that you 'have time to correct problems and so that structural changes are still relatively easy to make'. Issues of design, such as ensuring that stand-alone materials are 'written for a lower common denominator of student ability' (Nichols 1997: 370), should clearly have been taken into account prior to testing but may be more apparent after it! As mentioned above, tests may take a variety of forms but whether feedback is written or verbal, students should be free to do hands-on testing without your influence or interference. In terms of categorizing the results of your tests, a useful typology might follow that of Table 9.1.

Table 9.1 Fixing strategies

Level of Problem	Reason
1. Must fix	where users get stuck and can't recover
2. Should fix	where users may get stuck but usually manage to recover
3. Would be nice to fix	not a big cause of problems but diminishes overall quality
4. Fix when everything else is done	no one would really notice but we know it is there

Source: Boling and Frick 1997

Such a categorization allows some meaning to be applied to what may turn out to be a surprising number of 'glitches'. As we have stressed elsewhere, the temptation to get obsessed with minute details can be great but should be avoided. Ask 'is it useful?' not 'is it perfect?' since the latter is at the end of the rainbow. Avoid becoming a 'test junkie' (Boling and Frick 1997). Thinking about your testing in these terms should be helpful in this regard. Not getting obsessive could also be said to apply to testing itself; if perfection of an almost infinitely modifiable and changing resource is sought then testing could stretch to an impractical number of rounds. Observe the old Vulcan proverb: 'By accepting the inevitable one can achieve inner peace.' Our own experiences support those of Ward and Newlands (1998) that students express a clear preference for functionality over style. That said, strive for accuracy of content and structure, as you would using more traditional delivery systems. Correct obvious errors and mistakes. The last thing you want is your WBI to be held up to ridicule in an international forum. One site at the College of Education at Illinois State University was voted the 'worst dressed web course' because it was, among other things according to Boshier *et al.* (1997: 347), 'only marginally literate and replete with errors'. This type of notoriety will certainly have any hit counter you might install on your site spinning vigorously, but for the wrong reasons.

In stressing the linked nature of some stages of the process, some testing will have to take place 'live'. This means that testing prior to use needs to be followed up once the web pages are in actual use by students. Thus an email 'technical support hotline' is needed in the early stages of the project. This should be checked regularly, and students assured that action is being taken and problems categorized and acted upon according to a typology similar to the one outlined above. Electronic feedback forms can also be included as part of the project, using features such as radio buttons, checkboxes and text entry points; these can allow for comments on specific aspects of the project and allow rapid action to correct either the technical or student problem. Unlike feedback sought at the end of a programme of study (which must be comprehensive and can therefore be lengthy and discourage considered responses), such feedback can be brief and specific to a particular point in time/space. The greater likelihood of getting something done to affect your study will, it has been found, lead to higher and more interested responses.

Adapting

In a sense, it is important not to get too hung up over the issue of adaptation. If one is writing a book, even where this is 'only' support documentation for students, then the aim will be to make the product as perfect as possible and take seriously every infinitesimal error in spelling, grammar, referencing and so on. However, in setting up ODL pages on the web the goals are quite different from traditional teaching. Rather than concentrating on individual performance measures, ODL web-based learning involves complex situations of multiple variables and causality. Though the key challenge may be, in this case, to ensure that projects are sufficiently dynamic and active (Ravitz 1997), there can never be a 'finished' product in the sense that a book has to be, sooner or later, ready for publication. Furthermore, as alluded to earlier, web-based ODL materials may be used in ways that are not envisaged by a developer, or if envisaged, then given the exploratory nature of the web, hardly predictable in any precise way.

In this case, adapting your pages really does become adapting, rather than finishing. It is never possible to perfect the pages you are working on and, in this context, the crucial questions concern prioritizing which changes to make when, rather than which changes to make in order definitively to finish. Here, it has been stressed earlier that setting up WBI pages is not a short cut to less work. Studies tend to show that academics involved with such projects indicate an increase in both work and interactivity with

students (Cornell 1999). We can only support these findings.
Bearing this in mind, a practical approach needs to be taken so that
such projects do not become the equivalent of a bottomless pit into
which all efforts are expended. Thus adaptation becomes an
ongoing and measured process in which one does what one can.
Ease of adapting, updating and revising must also be considered at
the outset so that the processes of improvement and change can be
as painless as possible (Voithofer 1997). Adapting is intermingled
with other areas of work and with the various evaluative
techniques you will seek to employ such as transcript/content
analysis (Levin *et al.* 1990), portfolio review (CRESST 1996),
research methods for evaluating academic networked services
(McClure and Lopata 1996) and online survey research (Babbie
1990). These methods will help to interpret user testimonials,
volume of activity statistics, log-books, task phase analysis,
surveys and interviews, unsolicited feedback, case studies,
observation and expert reviews and so allow adaptation. What this
basically amounts to saying is that within the growth of your
pages, there are likely to be stages of development. It can be useful
to measure intermediate outcomes and to regard adaptation,
evaluation and testing as an integrated process which ultimately —
though you may start by testing, adapting and evaluating (in *that*
order) — comes in no particular order (Kozma and Quellmalz 1996).

The nuts and bolts of testing have not been described here in detail
since it might be difficult to account for all the variables that may
be encountered. Instead, what has been emphasized are two key
points. First, that testing and adaptation are absolutely critical.
Second, that neither should come to dominate your work. As
Trilling and Hood (1999: 17) put it, 'we need to go outside, breathe
deep, take a walk, smell the flowers and forget about technology at
least once a day'. Most of the hard work has, indeed, been done by

the time you are ready to test, and you must not let the lack of perfection in your product prevent the product being useful to students, nor lead you to neglect other aspects of your teaching/ research activities. Finally, one obvious way to adapt your own pages is to keep abreast of developments in the field and to look at the developments made by friends and colleagues at different institutions.

Evaluation

There are various aspects of evaluation to consider within WBI. As has already been mentioned, there is a need at the very earliest of the many planning stages you will go through to consider online evaluation. Some type of feedback is essential (Falk 1997; Stefanov *et al.* 1998). There are a variety of things you might want feedback on—ranging from design considerations through to the nature of the material itself (Graham and McNeil 1999; Halpin 1996; Rowntree 1998b). There are several ways in which you might glean this information. You will also want to evaluate learning outcomes (Arnone and Small 1999; Martin 1997). Again, how you do this will depend on the subject area, how it fits in with the course or module, year of study or personal preference. Finally, after committing considerable resources to the product you will almost certainly wish to do some kind of cost benefit analysis, for your own purposes and those of colleagues wishing to develop their own WBI. Here we discuss these in light of our own experiences.

Feedback

One big advantage of the WBI over more conventional methods of instruction is the ability to get instant, electronic feedback. By means of carefully crafted feedback forms students can be encouraged to provide teachers and designers with comments and information relating to various aspects of the site. Students' thoughts on content, layout, navigation, added functionality and the like can be gauged by means of feedback forms which require little time and effort to complete. Figure 9.1 shows the first seven elements of the feedback form for the *Social Geography and Nottingham* WBI. As you can see, although the form is based on features, such as radio buttons and pop-down menus, which are popular with the students, we do ask at the outset that they take some time to consider the questions. A further feature to encourage full and open contributions is the option to submit the form anonymously. This does result in some useful critical comment on the material as well as the standard, useless, predictable, banal witticisms common to any anonymous feedback.

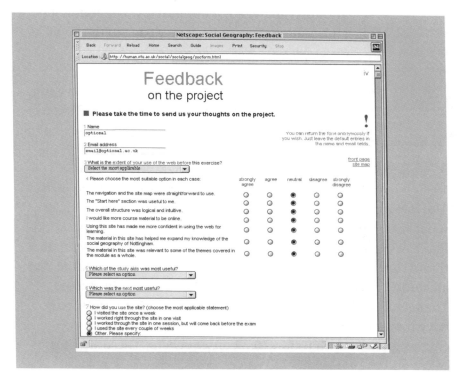

figure 9.1
Social Geography and Nottingham feedback form

How the feedback process takes place is dependent on a number of criteria:

- The personal preferences of the teacher. Do you want a formalized method of getting student feedback or do you prefer more informal channels?
- The nature of the material. Does your project lend itself to rapid response to feedback?
- What stage in the development process the WBI has reached. How much do you want to know about design and navigation considerations compared to problems with actual content?

One advantage of using web-based forms for assessment is that the results can readily be loaded into a spreadsheet and then analyzed quantitatively. The following tables show the results of feedback to the *Social Geography and Nottingham* site.

> One hundred per cent of Nottingham Trent University second year Geography students claimed some level of expertise on the web.
> SOCIAL GEOGRAPHY AND NOTTINGHAM
> FEEDBACK 1998/9

It was considered important at the outset to gauge student familiarity with the web. Table 9.2 shows the responses to the question 'What is the extent of your use of the web before this exercise?'. Of the 36 students taking the level two social geography module in 1998/9 more than half had reasonable familiarity with the web. The fact that 75 per cent had made moderate use of the web is what you might expect of students at this stage. As the use of the internet expands at home and in schools we should expect that figure to increase. However, it would be a mistake to assume that all students will be web-literate when they enter universities. There will be circumstances where conventional students may have had some exposure to the medium but in an unhappy environment or with unforeseen consequences. These students might be antipathetic towards the web and might require some special tuition. Students coming via less conventional routes might have little or no exposure to the internet and might also require special consideration at the early stages of exposure to WBI.

Table 9.2 Familiarity with the web,
Social Geography students 1998/9

Option	Response %
I have had no experience of the web	0
I have had limited experience of the web	28
I use the web quite often	50
I use the web regularly	22
I live online	0

It is also useful to determine the effectiveness of different aspects of the site. Table 9.3 shows how students responded to a variety of questions relating to navigation, structure, course material and the like. It would appear that this class was an agreeable bunch, at least as far as responses to these questions was concerned. Most found the navigation easy to use. Most found the online help useful. They were perhaps less enthusiastic about the overall structure of the site. They were very keen to see more course material on the site and this is something we are working towards, though this is the area where most problems of copyright occur. Just over half thought that they had gained confidence using the web. Since this WBI is largely self-contained and makes little use of links and search engines perhaps this is not an unsurprising result. All agreed that the material in the WBI had helped their understanding of the local social geography and of some of the themes within the discipline of social geography.

Table 9.3 Percentage response to content and structure,
Social Geography students 1998/9

Question	Strongly agree	Agree	Neutral	Disagree	Strongly disagree
The navigation was straightforward to use	33	50	9	8	0
The online help was useful to me	20	58	16	6	0
The overall structure was logical and intuitive	17	50	25	8	0
I would like more course material to be online	40	22	25	13	0
Using this site has made me more confident in using the web for learning	16	42	42	0	0
The material in this site has helped me expand my knowledge of the social geography of Nottingham	41	59	0	0	0
The material in this site was relevant to some of the themes covered in the module as a whole	9	91	0	0	0

Finding out how students made use of the site can also be of
interest. With the *Social Geography and Nottingham* material
students were encouraged to make use of the third of the allotted
three hours for the module during the last seven weeks during
which it ran. This was because it was integrated into the module
and was designed to follow on from the more formal classroom-

based instruction. Also, the tutor would be available if any problems or queries arose. However, because one of the main underlying reasons for constructing the site was to promote open and distance learning there was no compulsion for students to use that hour and only a handful made regular use of that particular period. Others were more flexible and imaginative (depending on your viewpoint) in their use of the site. Table 9.4 gives a breakdown of the use of the site. Of those offering other explanations some admitted to simply printing off what they saw as the relevant information — presumably to cram the night before the examination. One character professed a need to visit the site when he/she did not know what else to do! Evidently some students are still not quite sure how to interact with electronic media. While printing some relevant information, particularly where an examination is looming, may be desirable and expedient on the part of the student — and while this practice shifts costs from faculty to student — it largely renders the exercise of creating WBI projects nugatory.

Table 9.4 Students' use of the site, Social Geography students 1998/9

Option	Response %
I visited the site once a week	28
I worked right through the site in one visit	20
I worked right through in one session but I will return before the exam	11
I used the site every couple of weeks	36
Other	5

After spending a great deal of time and effort developing WBI it is useful to know how the site could be improved. In the *Social Geography and Nottingham* feedback form the following question was asked, related to a list of tickboxes: 'Please indicate what features you feel would improve the project'. Table 9.5 shows that despite the interactivity already available on the site, students would prefer more. More articles online are also a popular improvement. Admittedly there is a gap here and we would like to be able to post relevant articles or book chapters, or even key paragraphs from these. However, as Chapter 4 shows, the thorny problem of copyright means that the much-vaunted information superhighway will never be anything of the kind until government ensures that copyright laws protect authors but do not impede pedagogical progress. Students were also keen to have more email mentoring. This is a worthy aim and there are ways to do this that do not impinge unduly on tutors' time, either through notice-board discussion of a topic or by simply leaving aside a period of time each week to respond to individual student email. However, although around 15 per cent of the class thought email mentoring would improve the site, only a few individuals communicated with the tutor during the running of the module. A number suggested that faster download would be an improvement, but this is largely outwith your control in terms of hardware. However, if there are problems with download time it might be worthwhile revising the site design, particularly in terms of memory-heavy graphic elements. The other suggestions were not especially productive, although one enterprising and hopeful individual suggested that the posting of the forthcoming examination questions would be a good idea.

Table 9.5 Suggestions for improvements to the site, Social Geography students 1998/9

Option	Response %
Email mentoring (where you can email a tutor or expert)	15
More interactivity (things for you to do)	30
Assessed exercises (submitted for marking)	15
More articles	20
Faster download	15
Other	5

This gives some example of the range of feedback which you might use. Like any question and answer exercise there is a need to balance your own interests and requirements with those of the respondent. What may seem pertinent and clear to the instructor/designer may seem irrelevant or unclear to the student/end-user. We have found that some of the more significant and important feedback can simply be had in class or via email. Clearly, the more reliance there is on distance learning in your WBI, the greater will be the need to develop online and even interactive modes of feedback.

Assessing learning outcomes

This is another, and most important, aspect of evaluation. Was it worth it as far as the student is concerned? Assessing learner outcomes usually takes the form of a series of data and information points, collected during a course of study (Hudspeth 1997). There are a variety of learning outcomes that teachers would expect students to experience or achieve in any learning experience, and so many expectations that the individual student may have that assessing learning outcomes is one of the most problematic features of education. The learning outcomes in open and distance learning are particularly difficult to assess. Any exercise in assessing learning outcomes will depend on:

- How the WBI is integrated into the course, module, etc.
- Whether the WBI is a pilot project or designed as a full-scale teaching and learning experience.
- Whether the WBI, or aspects thereof, are to be formally assessed.

Lecter:

'A census taker tried to quantify me once. I ate his liver with some fava beans and a big Amarone. Go back to school, little Starling.'

Harris 1991: 22-23

It is a matter of personal preference how much of the assessment is directly incorporated into the WBI, what format that takes and how the student is expected to approach this. Some may choose to incorporate unassessed tests, quizzes and the like, within the site to retain students' interest and allow them to measure their own progress; others may like to incorporate the material delivered via the web into more conventional and formal assessment structures; yet others may prefer to mix the types of assessment or devise entirely new ones. The *Social Geography and Nottingham* WBI contains a test and a quiz, and it was made clear at the outset of the module that two out of six of the optional examination questions, such as these, would be based on the WBI:

Why and in what ways does crime vary over space? Illustrate your answer with particular reference to Nottingham.

With particular reference to Nottingham, discuss the social geography of housing and health.

There are a variety of ways that tests and quizzes can be implemented within your pages. Figure 9.2 shows how features, such as radio buttons and text areas, can be incorporated into a quiz—here, the first two questions from the *Social Geography and Nottingham* 20-question quiz. This was generated using the CASTLE Toolkit and based on the thematic section of the WBI. The results are generated automatically so that the student can check progress. A different approach was used to test students' familiarity with the section devoted to the different areas of the city. Here, 28 questions are provided to gauge students' own familiarity with the city, as well as material contained within the WBI. Questions are asked of a series of photographs and users are asked to respond to one of the questions (Figure 9.3). Clicking option a) returns a response of 'Possibly. But not' and invites the student to click on 'Try Again'; b) returns 'Not in The Park' (the better-off area in which the house is situated), and again the user may 'Try Again'; c) returns 'Yes. A really user-friendly fire escape' and invites the user to attempt the 'Next question'. Although there is humour in some of the questions, they have proved a valuable tool for student revision. Indeed, do not be afraid to include humour in your pages. This helps to keep the students interested. However, avoid excess frivolity, otherwise the students might fail to take your WBI seriously. Of the 20 questions in the *Social Geography and Nottingham* quiz only one contains a frivolous element (Figure 9.4). But, even here, the correct answer is among those offered in the list.

The *e:net* WBI includes a specially prepared and relevant Environment in International Relations Quiz and also references other quizzes already available on web pages, and on a wide variety of topics from nuclear issues and toxic waste to compost preparation! It was emphasized at the outset that the information

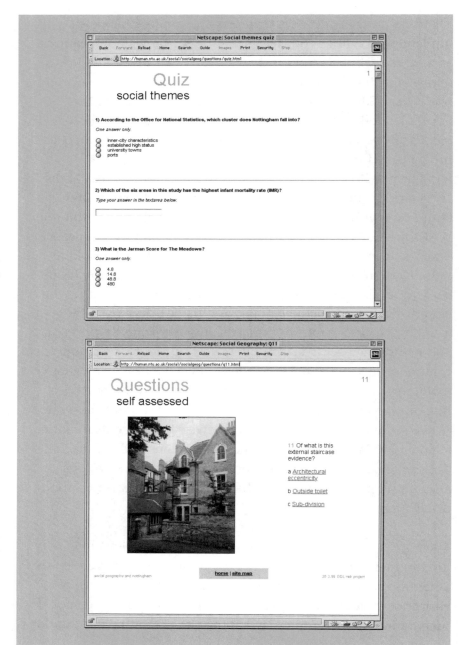

figure 9.2
*Social
Geography
and
Nottingham*
quiz

figure 9.3
*Social
Geography
and
Nottingham*
self
assessed
test

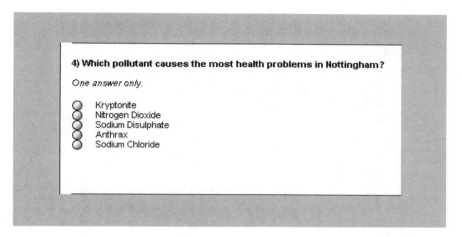

4) Which pollutant causes the most health problems in Nottingham?

One answer only.

○ Kryptonite
○ Nitrogen Dioxide
○ Sodium Disulphate
○ Anthrax
○ Sodium Chloride

figure 9.4
introduce
humour
into your
site

offered on *e:net* would be ideal in helping to answer essay and exam questions. Sample questions were included for each topic and subtopic so that the students could see clearly how to integrate the pages' empirical data into their theoretical arguments. In the *Medieval History* site no formal assessment is included as part of the WBI. Neither is there any stand-alone informal assessment, at the time of writing. The WBI takes the form of a study companion, to travel alongside the students as they progress through the module. Each section is tied in to the seminar work for that week and includes instructions for the preparation for that week, advice on how to approach it, references and directions to information sources. Where the seminar session will incorporate an assessed element (group presentations and also compulsory directed learning exercises), the rubric, material and guidelines for this are also included (Figure 9.5).

It would also be advantageous and worthwhile to determine to what extent the use of the web as a medium contributed to the learning experience of

'At some point in time there must be an assessment of the extent to which the technology has added to the learning accomplished.'

BUTLER 1999: 64

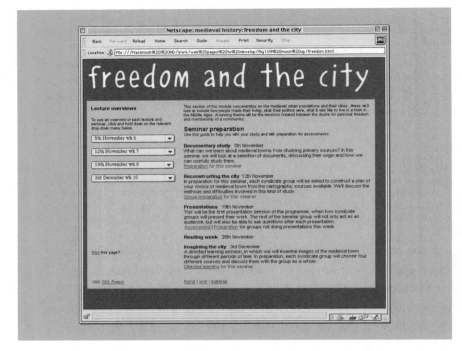

figure 9.5
Medieval History: information about assessment and preparation

the students (Dillon and Zhu 1997; Reeves and Reeves 1997). Could the material have been delivered by more conventional means? Did web delivery add anything to the learning experience? While the material in Tables 9.2 to 9.6 reveal some interesting insights, the questions contained therein were designed as much to allow development of the site as explore the learning experience of web delivery as such. However, while we can ask whether students enjoyed using a site and got a valuable learning experience by using the site, it is very difficult to devise methodologies to measure the value added. 'The time-honored experimental treatments which require control groups and different treatments are,' in the view of Butler (1999: 64), 'difficult to translate to assessment of the value added by the Internet in creating a learning environment.'

In response to the question 'Which of the study aids was the most useful?' students appeared to have clear ideas of their most favoured study aid as Table 9.6 shows.

Table 9.6 Responses to utility of study aids, Social Geography students 1998/9

Option	Most useful %	Next most useful %
Self-assessed quiz	26	50
Scored test	38	10
Sample exam questions	21	32
Advice in the 'start here' section	15	8

Cost benefit analysis

One final aspect of the assessment process, and the main thing to consider at the outset, is how the costs in terms of time, funds and effort are to be balanced by the benefits. There is no doubt that our students have made good use of the three WBI products which we have devised. There have been clear learning outcomes and students have generally been enthusiastic and sympathetic towards innovative learning and teaching formats — they approve. However much is gained in terms of approbation from students and colleagues, the level of investment very often shocks newcomers to WBI. There are no short cuts and a considerable amount of time and effort is needed. Unless you work in a very generous and far-sighted institution that can allow you a sabbatical or some type of relief to pursue your plans much of this time will be your own, even if you have technical and other support. The *Social Geography and Nottingham* WBI Mark I developed out of our own interests and was the product of our own time. However, due to the generosity of The Nottingham Trent University we were able to employ an assistant dedicated to working on the three projects. The funds allowed some relief from teaching — the saved time being devoted to the projects. We have also been able to purchase much needed hardware and software. This allowed us to develop the *Social Geography and Nottingham* WBI and initiate both the *e:net* and the *Medieval History* projects. You can do some good work, then, within your own parameters, but so much more can be achieved with financial and technical help.

Although it might be possible to create an online course in a day (Polyson 1997), you will find that many, many days will be required to create something you can be proud of, that colleagues will envy and that the students will be satisfied with. Darby (1999) has researched the value of ODL via the internet and concluded that the economics of WBI are such that it is only viable with classes of 100 to 200. Ever-increasing class sizes and increasing demands on lecturers' time for research and administrative duties make WBI a more attractive option for institutions trying to keep ahead of the field in an increasingly competitive market. But institutional management must realize that there needs to be a commitment to staff training and infrastructural investment (Hill 1997; Jennings and Dirksen 1997; Rossner and Stockley 1997). The ideal scenario is a team approach, where at least one member of the team has technical knowledge of the web (platforms, servers, and the like) and is familiar with HTML and other relevant software. The best sites are *instructed* by individuals possessing certain skills and *produced* by individuals with another set of skills (Boshier *et al.* 1997). Ideally, then, there should be close cooperation between educators and technical professionals (Gray 1997; Luchini

1998; Maddux *et al.* 1999; Trilling and Hood 1999; Willis and Dickinson 1997; Wilson 1999). This should entail the employment of skilled support staff who are employed on the same status as lecturers, otherwise they will end up in the subservient and skivvy role that far too many academics ascribe to support staff, and undue conflict is likely to result. Unless a 'non-academic' member of the team has equality of status then scenarios can be envisaged in which that member will be overburdened by collecting material and other chores that should be those of the 'traditional' members of the team. However, this is likely to meet the type of academic stonewalling described in Chapter 5. There should also be an institutional commitment to train existing teaching staff to construct and maintain their own WBI. This must go further than a familiarity with the technology. The acquisition of technical skill is no guarantee of efficient use (Herrmann *et al.* 1999).

> 'University faculty members are the logical choice to provide subject matter, expertise and instructional design. However, university faculty should not be expected to complete the technical work of building the courseware because they must meet numerous other performance measures, and may lack critical technical skills.'
>
> BRAHLER *ET AL*. 1999: 52

In an extensive review of the costs of developing online materials for higher education, Brahler *et al.* (1999) show that calculating costs is very difficult but conclude that developing online materials is very costly. They show that development times for computer assisted instructional materials can range from 50 to 350 hours for each hour of teaching time. The time costs of design and development in WBI are much higher than for traditional instructional methods (Williams and Peters 1997). However, costs decrease with developer experience, but will rise with the ambitiousness of the project.

The total number of hours spent on planning, constructing, implementing, testing, adapting and evaluating the *Social*

Geography and Nottingham WBI is difficult to gauge, but estimates suggest that in excess of 300 hours have been spent on the project. This, in accountants' terms, is a poor investment for the seven hours or so of formal teaching time 'freed up' during term. However, since this was the pilot, there was a learning process to account for and much catching up to do. Hence this book, which we believe will provide a time-saving guide to avoiding the pitfalls and culs-de-sac which cost us dear. Although costs appear to be high, the 'free time' aspect was never a consideration in the accounting. And, as explained in Chapter 5, any time 'gained' in formal teaching is absorbed in developing the site or fine-tuning the existing pages. No, the benefits are difficult to quantify. There is the personal satisfaction of having created something new and of having acquired new skills in so doing. There is a certain amount of kudos to be gained. *Curricula vitae* have been enhanced. But, most important of all, the teaching and learning experiences of staff and students have been enriched.

E:net took some considerable time to set up in terms of researching and designing the pages, though it is difficult to tell exactly how many hours. Fewer than the *Social Geography and Nottingham* site, certainly. The nature of the module, as linked to the research interests of the module leader, meant that some of the searching for information did, in any case, form part of the work of the lecturer. Furthermore, once the decision had been taken to launch *e:net*, information was gathered in incremental fashion as part of the research process and put on one side. Even so, directed efforts have been required to get *e:net* off the ground in terms of designing and organizing material; whilst these efforts should not be underestimated in terms of the time they took up, they provide a solid basis for independent student research and this in turn has fed insights back into the lecturer's own research. The previously wasted effort involved in students and lecturers mistyping URLs and so on has been avoided, as has the frustration of students unable to get hold of limited library resources.

If it is difficult to generate a cost benefit analysis of the longer-established WBIs, it is more problematic doing so for the newest product. Nevertheless, around 40 hours have been devoted to the construction and testing of the *Medieval History* site. Gains will be in staff and student time. This will be limited and mostly in the form of students being able to access module-specific information online without spending time hunting down staff and staff having to rifle around for said information. There will be gains in reprographic costs, though since students like to have a hard copy of everything costs are simply defrayed to Computing Services or the students. Students have been favourable thus far, especially the mature students and others living off campus who can access the material from the comfort of their own homes, which may be some distance away.

There are, of course, students' costs to consider. For students to access your pages from home implies certain standards of hardware and software. These will usually have to be borne by the students themselves at a time when they are under increasing financial strain. Because of this increasing burden, more students are living off campus in the parental home and more are attending part-time (either officially registered on a part-time course or else in reality, since increasing numbers are working hours which officially put them into the realm of full-time employment). For these, and the growing number of mature students, distance WBI will become a more attractive option. They will make their own cost benefit analysis. As well as student financial costs it is also worth considering their time costs. These can often be perceived as being high, particularly among relatively untrained humanities and social science students (Usip and Bee 1998). However, if you follow our advice about integrating basic training and introductory sessions into your WBI these costs should be minimized.

Costs common to the three WBIs include:
- Time—this cannot be avoided; the more complex the site, the more time involved and this valuable commodity will have to be found. Try to get help.
- Effort—this cannot be avoided. Getting help might spread the load.
- Money—new software might be required as well as updates of existing software, new hardware such as scanners, digital cameras might have to be purchased, but could be borrowed.

Benefits common to the three WBIs include:
- Time—but savings are minimal and would have accountants weeping onto their spreadsheets.
- Flexibility for you—you can update as you go, unlike paper handouts which are costly to reproduce during term.
- Flexibility for the student—they can access the material remotely and at three in the morning, if they wish.
- You are forced to rethink basic pedagogical issues. For example, Marina Orsini-Jones and Anne Davidson (1999: 37) found that they had 'a unique opportunity to explore pedagogic issues relating to educational technology in an interdisciplinary way', when working on a collaborative web project. This has been our experience also.

We hope you are encouraged by this book to use the web in your teaching. We make no secret that there is a lot of work involved—

as in any teaching and learning experience—and accept that some sections might be discouraging to some. Similarly there are many rewards—personal and pedagogical. Using WBI will not dramatically alter the way you teach, but it will add to your skills base and give an added dimension to your teaching. It does take a considerable amount of time, effort and commitment to produce a meaningful teaching and learning product using your own WBI and at times you may feel, like Whitman, that you are giving yourself, but by persevering you will be taking the technology of the twenty-first century further for yourself and for your students.

glossary

glossary

Applet

Small Java program that can be embedded into a web page to produce, among other things, animations, activities or calculations.

Asynchronous conferencing

Communication involving multiple people that does not take place in real time. Examples include message boards and mailing lists.

Bandwidth

The amount of data that can be transferred through your internet connection. Effectively, this means the speed at which data will download. It is limited by the slowest link between you and the data, even if your own machine has a very fast connection. *Low bandwidth* is often used to denote web sites that are made up of pages and graphics with low file sizes and which therefore load quickly even through a slow connection.

Baud

A measurement of data transmission speed, often used interchangeably with *bits per second*, although they are not the same thing. In practice, the rate at which a modem sends and receives information.

Block

In a document hierarchy, elements that exist below the level of a page, but above individual words — such as a paragraph.

Browser

The software that lets you view web pages, by translating HTML script and embedded elements into text, graphics and so forth. The most popular browsers at time of writing are *Netscape Navigator* and *Microsoft Internet Explorer*.

Byte

A unit of computer storage or memory and an abbreviation of *binary term*. One byte is very tiny and larger quantities of memory are expressed as kilobytes (K), megabytes (Mb) or gigabytes (Gb).

CAL

An abbreviation of *computer aided learning*, CAL is learning delivered through IT, particularly in the form of structured, directed modules. A piece of such learning software may be referred to as 'a CAL'.

CGI
An abbreviation of *Common Gateway Interface*. A CGI is usually a small program that receives information from a web page and does something with it. For example, it may be used to put the content of a form into an email message and send the message on.

Click stream
Often used to denote the average or most popular routes taken through a web site.

Conferencing
See *Asynchronous* and *Synchronous* conferencing.

CSS
An abbreviation of *Cascading Style Sheets*.

Degradable
In terms of the web, pages that take into account compatibility problems related to successive upgrades of browser software. A page that degrades gracefully is one that works well in older browsers.

Download
To request and receive data via a network or the internet.

dpi
An abbreviation of *dots per inch*, dpi denotes the resolution of screen or print. The higher the dpi, the better quality the resolution. The average monitor resolution is 72 dpi.

Explorer
Microsoft's web browser.

Frames
Subdivided sections of the web browser window that act as different pages. The user sees one page divided into two or more parts, which can then act independently, or influence each other.

Freeware

Software that is provided free, often by independent developers. Related to *Careware*, the licensing for which may stipulate a nominal toll—for example, that you send a postcard to the developer.

GIF

An abbreviation for graphics interchange format, GIF uses 'lossless' compression to render images with small file sizes, but no loss in quality. A common format for web graphics.

Hit

A visit to a web page.

HTML

Hypertext Mark-up Language is the script used to write pages for the web.

http

An abbreviation for *hypertext transfer* (transport, transmission) *protocol*, the means by which computers transmit data for the web.

Hyperlink

A method of linking from one web page to another, a hyperlink can take the form of text or graphic.

Inline

In a document hierarchy, elements that exist below the level of a *block*—such as an individual word.

Interface

In terms of web pages, the environment which mediates between the user and the functions and contents of the site.

Internet Service Provider (ISP)

Commercial supplier of access to the internet for individuals.

Internet relay chat (IRC)

A widespread system for real-time, text-based conversation between two or more people over the internet.

Java
A compiled programming language, developed by Sun Microsystems.

JavaScript
An interpreted scripting language, JavaScript can be incorporated directly into a web page and is widely used to add interactivity.

JPG, JPEG
An abbreviation for Joint Photographic Experts Group, JPEG compression reduces file size by discarding colour information (lossy compression). A common format for web graphics.

K, k
A unit of computer storage or memory, 1K or kilobyte, equals a thousand (1024) bytes.

Link
See *Hyperlink*.

Mb
A unit of computer storage or memory, 1Mb or Megabyte, equals a thousand kilobytes.

Mailing list
In terms of email, equivalent to a postal mailing list, with the exception that the individual members can email back to the list and all members can see all messages.

Mean paths
The most frequented routes through a web site.

Message board
Widespread facility for text-based, non-real-time messaging and conversation, message boards are usually arranged into conversational topics or threads.

Multimedia
In terms of the web, multimedia is usually used to denote pages that incorporate animated images, video or sound.

Navigation
The means or signposts by which the user is directed, or invited, to move around pages within a site.

Netscape
The common reference for Netscape Corporation's web browser.

Newsgroup
A topic-based discussion group on Usenet.

Online discussion
This often refers to text-based discussion, although voice-based conversation is possible with the correct equipment.

PDF
An abbreviation for Portable Document Format, PDF is a cross-platform form of document delivery developed by Adobe Inc.

Path
In terms of the web, the route a user takes through a site.

Platform
A combination of a computer's hardware and Operating System. Two major platforms are Wintel (Windows software and Intel processor) and Mac.

Plug-in
An add-on piece of software that extends the browser's capability to handle a specific feature embedded in web pages.

Resolution
See *dpi*.

Sans-serif
One of the two main categories of typeface. Sans-serif typefaces are simpler than serif, and usually do not have hooks and feet on the ascenders and descenders of letters. An example is Helvetica.

Search engine

A facility that allows the user to search the web using keywords. Examples include Yahoo, Excite and Alta Vista.

Screen reader

Software that translates text on screen to an aural reading.

Screen resolution

See *dpi*.

Serif

One of the two main categories of typeface. Serif typefaces have hooks and feet on the ascenders and descenders of letters and thick and thin parts. An example is Times.

Server

In terms of the web, a combination of hardware and software which allows access to the site and on which the web pages are physically located.

Server administrator

The person who manages the server and often the web site.

Session depth

A combination of the time a user spends at a site and the number of pages he or she visits.

Shareware

Software that is provided for a low or nominal fee, often by independent developers.

Synchronous conferencing

Communication involving multiple people that takes place in real time. Can be conducted by text or voice. An example is the chat rooms that proliferate on the internet.

Tag

Fundamental part of the syntax of HTML.

Traffic
Often, the number of people logged on to a network at once, or the amount of activity on a site at one time.

Upload
To send data via a network or the internet.

URL
An abbreviation for *Uniform (*Universal*) Resource Locator*, URL is, in practice, the address of a web page or site.

Web browser
See *browser*.

Wysiwyg
An abbreviation for *what you see is what you get*. In terms of web editing software, WYSIWYG refers to packages that allow the user to lay out a page without recourse to using HTML.

X-height
In a typeface, the height of the letters that have no ascenders or descenders, like x.

references
references

Air Force Distance Learning Office (1998) *Distance Learning Resource Handbook*, sixth edition, http://www.au.mil/afdlo/

Akers, R. (1997) Web discussion forums in teaching and learning, *Technology Source*, August, http://horizon.unc.edu/TS/cases/1997-08.asp

Allen, J. (1999) The writing on the web, *Webmonkey*, 2 August, http://www.hotwired.com/webmonkey/99/32/index0a.html

Arlidge, J. (1997) Headlines caught in tangled Nets of Shetland rivals, *The Observer*, 9 November, p. 6

Arnone, M.P. and Small, R.V. (1999) Evaluating the motivational effectiveness of children's websites, *Educational Technology*, **39**(2), 51–55

Babbie, E. (1990) *Survey Research Methods*, second edition, Wadsworth, Belmont, VA

Bannan, B. and Milheim, W.D. (1997) Existing web-based instruction courses and their design, in B.H. Khan (ed.) *Web-Based Instruction*, Educational Technology Publications, Englewood Cliffs, NJ, pp. 381–387

Barker, P. (1998a) Using intranets to support teaching and learning, *IETI*, **36**(1), 3–9

Barker, P. (1998b) Virtual universities and electronic course delivery, Paper presented at the CRICS IV International Conference, San Jose, Costa Rica

Barron, A.E., Tompkins, B. and Tai, D. (1996) Design guidelines for the World Wide Web, *Journal of Interactive Instruction Development*, **8**(3), 13–17

Bates, A. (1997) The impact of technological change on open and distance learning, *Distance Education*, **18**(1), 93–109

Berge, Z.L. (1999) Interaction in post-secondary web-based learning, *Educational Technology*, **39**(1), 5–11

Bickford, P. (1999) The great internet fashion show, *View Source Magazine*, April,
http://developer.netscape.com/viewsource/bickford_fashion.html

Boling, E. and Frick, T. (1997) Holistic rapid prototyping for web design: early usability testing is essential, in B.H. Khan (ed.) *Web-Based Instruction*, Educational Technology Publications, Englewood Cliffs, NJ, pp. 319–328

Boshier, R., Mohapi, M., Moulton, G., Qayyum, A., Sadownik, L. and Wilson, M. (1997) Best and worst dressed web courses: strutting into the 21st century in comfort and style, *Distance Education*, **18**(2), 327–349

Bostock, S.J. (1997) Designing web-based instruction for active learning, in B.H. Khan (ed.) *Web-Based Instruction*, Educational Technology Publications, Englewood Cliffs, NJ, pp. 225–230

Bowers, C. (1998) The paradox of technology: what's gained and lost, *Thought and Action*, **14**(1), 49–57

Boyd, W. (1998) *Armadillo*, Penguin, London.

Brahler, G.J., Peterson, N.S. and Johnson, E.C. (1999) Developing on-line learning materials for higher education: an overview of current issues, *Educational Technology and Society*, **2**(2), 42–54

Branscomb, A.W. (1994) *Who Owns Information?*, Basic Books, New York

Brooks, D. (1997) *Web-Teaching: A Guide to Designing Interactive Teaching for the World Wide Web*, Plenum Press, New York

Browell, S. (1997) Open learning and multimedia — the legal issues, *Open Learning*, **12**(1), 52–57

Business Post (1999) Search engine blocks 'a nightmare', *Nottingham Evening Post*, 20 April

Butler, B.S. (1997) Using the World Wide Web to support classroom-based education: conclusions from a multiple-case study, in B.H. Khan (ed.) *Web-Based Instruction*, Educational Technology Publications, Englewood Cliffs, NJ, pp. 417–430

Butler, J.C. (1999) The tangled Web, *Active Learning*, **10**, 63–66

Cabell, B., Rencis, J.J., Grandin, H.T. and Alam, J. (1997) Using Java to develop interactive learning material for the World Wide Web, *International Journal of Engineering Education*, **13**(6), 397–406

Canadian Telework Association (1998) Distance education, Nepean, Ontario, http://www.ivc.ca/part10.html

Carter, V. (1996) Do media influence learning: revisiting the debate in the context of distance education, *Open Learning*, **11**(1), 31–40

Carty, S.K. (1998) The challenge of the educated web, *Searcher*, **6**(4), 32–33, 36–40

Center for Applied Special Technology (CAST) (1999) *Bobby 3.1.1*, http://www.cast.org/bobby/ (6.9.99 edition)

CRESST (Center for Research on Evaluation Standards and Student Testing) (1996) Creating Better Student Assessments, in *Improving America's Schools: A Newsletter on Issues in Social Reform*, http://www.ed.gov/pubs/IASA/newsletters/

Charlesworth, A. (1994) Tooling up on the law, *Times Higher Education Supplement, Multimedia*, 5, p. xiv

Charlesworth, A. (1995) Legal issues of WWW and electronic publishing, Paper presented at the Support Initiative for Multimedia Applications Workshop at Loughborough University February 13–14, http://www.man.ac.uk/MVC/SIMA/WWW/legal.html

Chen, D. and Zhao, Y. (1997) The web and homepage maker: making it easier to develop content on the WWW, *Computer Assisted Language Learning*, **10**(5), 427–441

Clark, R.E. and Estes, F. (1999) The development of authentic educational technologies, *Educational Technology*, **39**(2), 5–16

Coleman, M. (1999) Electronics education through integrated learning systems, *Electronics Education*, Summer, 29–31

Collis, B. and Smith, C. (1997) Desktop multimedia environments to support collaborative distance learning, *Instructional Science*, **25**(6), 433–462

Collis, B., Andernach, T. and Van Diepen, N. (1997) Web environments for group-based project work in higher education, *International Journal of Educational Telecommunications*, **3**(2–3), 109–130

Cornell, R. (1999) The onrush of technology in education: the professor's new dilemma, *Educational Technology*, **39**(3), 60–64

Cornell, R. and Martin, B.L. (1997) The role of motivation in web-based instruction, in B.H. Khan (ed.) *Web-Based Instruction*, Educational Technology Publications, Englewood Cliffs, NJ, pp.93–100

Crossman, D.M. (1997) The evolution of the World Wide Web as an emerging instructional technology tool, in B.H. Khan (ed.) *Web-Based Instruction*, Educational Technology Publications, Englewood Cliffs, NJ, pp. 19–23

Cummings, J.A. (1998) Promoting student interaction in the virtual college classroom, *Indiana Higher Education Faculty Papers*, http://www.ihets.org/distance_ed/fdpapers/1998/52.html

Dahmer, B. (1993) When technologies connect, *Training and Development (USA)*, **47**(1), 46–55

Daniel, J. (1996) *Mega-Universities and Knowledge Media: Technology Strategies for Higher Education*, Open University, Milton Keynes

Daniel, J. (1999) Distance learning in the era of networks: what are the key technologies?, *The Bulletin*, **138**, 7–9

Darby, J. (1999) The economics of open learning via the internet in the United Kingdom, *Proceedings of the PanCommonwealth Forum on Open Learning*, Brunei, March 1–5, http://www.col.org/forum/PCFpapers/PostWork/darby.pdf

Dick, W. and Reiser, R. (1989) *Planning Effective Instruction*, Prentice Hall, Englewood Cliffs, NJ

Dillon, A. and Zhu, E. (1997) Designing web-based instruction: a human-computer interaction perspective, in B.H. Khan (ed.) *Web-Based Instruction*, Educational Technology Publications, Englewood Cliffs, NJ, pp. 221–224

Doherty, A. (1998) The internet: destined to become a passive surfing technology, *Educational Technology*, **38**(5), 61–63

Douvains, G. (1997) Copyright law and distance learning technology: fair use in far classrooms, *International Journal of Instructional Media*, **24**(4), 43–46

Duchastel, P. (1997) A web-based model for university instruction, *Journal of Educational Technology Systems*, **25**(3), 221–228

Eco, U. (1983) *The Name of the Rose*, translation, Picador, London

Edwards, T. (1996) Community learning with an intelligent college: a new way to learn, *Journal of European Industrial Training*, **20**(1), 4–14

EFPICC (1999a) Leading European user and technology industry groups unite to call for a balanced Information Society Copyright Directive, EFPICC Information Letter, 7 July, http://www.eblida.org/lobby/position/efpiccle.htm

EFPICC (1999b) Consumers call on EU Ministers to allow lawful copying, EFPICC Press Release, 12 July, http://www.eblida.org/efpicc/efpiccp2.htm

EFPICC (1999c) The Consumer Fair Practice Campaign warns that information poverty could result from the new EU-Copyright proposal, EFPICC Press Release, 7 July, http://www.eblida.org/efpicc/press1.htm

Eklund, J. (1999) The role of student knowledge in the design of computer-based learning environments, *Educational Technology and Society*, **2**(1), 7–9

Enghagen, L.K. (1998) Copyright law and fair use: why ignorance is not bliss—a case for using guidelines, *Technology Source*, April, http://horizon.unc.edu/TS/commmentary/1998-04.asp

Falk, D.R. (1997) Designing a course on the World Wide Web, *Journal of Interactive Instruction Development*, **9**(4), 10–19

Firdyiwek, Y. (1999) Web-based courseware tools: where is the pedagogy?, *Educational Technology*, **39**(1), 29–34

Flake, J, (1996) The World Wide Web and Education, *Computers in Schools*, **12**(1–2), 89–100

Forinash, K., Rumsey, W. and Wisman, R. (1998) Interactive and collaborative uses of the web, *Indiana Higher Education Faculty Papers*, http://www.ihets.org/distance_ed/fdpapers/1998/46.html

Freedman, J. (1995) Using the World Wide Web to deliver educational software, *Multimedia Monitor*, **13**(11), 19–22

Frick, T.W., Corry, M. and Bray, M. (1997) Preparing and managing a course web site: understanding systemic change in education, in B.H. Khan (ed.) *Web-Based Instruction*, Educational Technology Publications, Englewood Cliffs, NJ, pp. 431–436

Friedrich, F. (1997) Transfer of learning technologies: the experience of the DELTA DEMO ESC project, *Open Learning*, **12**(3), 34–42

Gillani, B.B. and Relan, A. (1997) Incorporating interactivity and multimedia into web-based instruction, in B.H. Khan (ed.) *Web-Based Instruction*, Educational Technology Publications, Englewood Cliffs, NJ, pp. 231–237

Goldberg, M.W. (1997) Using a web-based course authoring tool to develop sophisticated web-based courses, in B.H. Khan (ed.) *Web-Based Instruction*, Educational Technology Publications, Englewood Cliffs, NJ, pp. 307–312

Graham, D.T. and McNeil, J. (1999) Using the internet as part of directed learning in social geography: developing web pages as an introduction to local social geography, *Journal of Geography in Higher Education*, **23**(2), 181–194

Graham, D.T., McNeil, J. and Pettiford, L. (2000) Problems and potentials of ODL: examples from NTU, *Innovations*, forthcoming

Gray, S. (1997) Training teachers, faculty members, and staff, in B.H. Khan (ed.) *Web-Based Instruction*, Educational Technology Publications, Englewood Cliffs, NJ, pp. 329–332

Greene, J. (1999) Active learning and open learning, *Proceedings of the PanCommonwealth Forum on Open Learning*, Brunei, March 1–5, http://www.col.org/forum/casestudies.htm

Halpin, R. (1996) Incorporating computer applications into inservice and preservice education: mathematics teachers explore the World Wide Web, *Journal of Technology and Teacher Education* **4**(3–4), 297–308

Hansen, L. and Frick, T.W. (1997) Evaluation guidelines for web-based course authoring systems, in B.H. Khan (ed.) *Web-Based Instruction*, Educational Technology Publications, Englewood Cliffs, NJ, pp. 299–306

Harmon, S. and Hurumi, A. (1996) A systematic approach to the integration of interactive distance learning into education and training, *Journal of Education for Business*, **71**(5), 267–271

Harris, M. (1998) Removing the 'distance' from distance learning: a perspective on student teacher interaction in the OMNIBUS program, *Michigan Community College Journal: Research and Practice*, **4**(1), 49–57

Harris, P., Harris, M. and Hannah, S. (1998) Confronting hypertext: exploring divergent responses to digital coursework, *Internet and Higher Education*, **1**(1), 45–57

Harris, T. (1991) *The Silence of the Lambs*, Mandarin, London

Harris, T. (1999) *Hannibal*, William Heinemann, London

Haugaard, E. (1976) *Hans Andersen: His Classic Fairy Tales*, translation, Victor Gollancz, London

Hedberg, J., Brown, C. and Arrighi, M. (1997) Interactive multimedia and Web-based learning: similarities and differences, in B.H. Khan (ed.) *Web-Based Instruction*, Educational Technology Publications, Englewood Cliffs, NJ, pp. 47–58

Herrmann, A., Fox, B. and Boyd, A. (1999) Using the World Wide Web in distance education programs in Australia, *Proceedings of the PanCommonwealth Forum on Open Learning*, Brunei, March 1–5, http://www.col.org/forum/casestudies.htm

Hill, J.R. (1997) Distance learning environments via the World Wide Web, in B.H. Khan (ed.) *Web-Based Instruction*, Educational Technology Publications, Englewood Cliffs, NJ, pp. 75–80

Hodes, C. (1998) Developing a rationale for technology integration, *Journal of Educational Technology Systems*, **26**(3), 225–234

Hudspeth, D. (1997) Testing learner outcomes in web-based instruction, in B.H. Khan (ed.) *Web-Based Instruction*, Educational Technology Publications, Englewood Cliffs, NJ, pp. 353–356

Innes, M.M. (1995) *The Metamorphoses of Ovid*, translation, Penguin, Harmondsworth

James, D.L. (1998) Distance learning, *Education-line*, http://www.leeds.ac.uk/educol/documents/000000142.htm

Jennings, M.M. and Dirksen, D.J. (1997) Facilitating change: a process for adoption and web-based instruction, in B.H. Khan (ed.) *Web-Based Instruction*, Educational Technology Publications, Englewood Cliffs, NJ, pp. 111–116

Jevons, F. and Northcott, P. (1994) *Costs and Quality in Resource Based Learning On-and-Off Campus*, NBEET Commissioned Report No.33: Canberra

Johnson, R. (1998) Web navigation: designing the user experience, *Mantex Newsletter*, Issue 1, December, http://www.mantex.co.uk/reviews/fleming.htm (19.7.99)

Kapur, S. and Stillman, G. (1997) Teaching and learning using the world wide web: a case study, *Innovation in Education and Training International*, **34**(4), 316–322

Kaufman, R. (1998) The internet as the ultimate technology and panacea, *Educational Technology*, **38**(1), 63–64

Kaur, A., Fadzil, M. and Baba, S. (1999) Design considerations for an on-line course: a case study in Malaysia, *Proceedings of the PanCommonwealth Forum on Open Learning*, Brunei, March 1–5, http://www.col.org/forum/casestudies.htm

Keady, Helen (1999) Using computer based materials on the undergraduate programme, *Innovation in Learning and Teaching*, **3**, 29–34

Kearsley, G. (1998) Educational technology: a critique, *Educational Technology*, **38**(2), 47–51

Kerven, D., Ambos, E. and Frost, E. (1998) Interactive web-based quizzes using the review automated generation system (RAGS), *International Journal of Educational Telecommunications*, **4**(1), 31–44

Kessell, S. (1999) Postgraduate courses on the WWW: teaching teachers and educating the professors, *Technology Source*, November, http://horizon.unc.edu/TS/development/1999-02.asp

Khan, B.H. (1997a) Web-based instruction (WBI): what is it and why is it?, in B.H. Khan (ed.) *Web-Based Instruction*, Educational Technology Publications, Englewood Cliffs, NJ, pp. 5–18

Khan, B.H. (ed.) (1997b) *Web-Based Instruction*, Educational Technology Publications, Englewood Cliffs, NJ

Khan, B.H. and Vega, R. (1997) Factors to consider when evaluating a web-based instruction course: a survey, in B.H. Khan (ed.) *Web-Based Instruction*, Educational Technology Publications, Englewood Cliffs, NJ, pp. 375–378

Klonoski, E. (1997) Studying technology using technology: an interdisciplinary course on the internet, *Technology Source*, September, http://horizon.unc.edu/TS/featured/1997-09.asp

Kolb, D.A. and Fry, R. (1975) Towards an applied theory of experiential learning, in C.L. Cooper, (ed.) *Theories of Group Processes*, John Wiley, London, pp. 33–57

Kozma, R. and Quellmalz, E. (1996) Issues and needs in evaluating the educational impacts of the national information infrastructure, Paper commissioned by the US Department of Education's Office of Educational Technology, http//www.ed.gov/Technology/

Leahy, C. (1999) Computer mediated communication, Paper presented at the Evolution conference on Information Technology in Humanities Learning and Teaching, Nottingham Trent University, 29 June

Lee, V., Murphy, D., Chan, C.C. and Chung, L. (1997) Computer aided distance learning: a case study, *Open Learning*, **12**(1), 58–62

Leu, D.J. and Leu, D.D. (1999) *Teaching with the Internet: Lessons from the Classroom*, Christopher-Gordon Publishers, Norwood, MA

Levin, B.H. (1998) Distance learning: technology and choices, *Report of the Blue Ridge Community College Office of Institutional Research*, Weyers Cave, VA

Levin, J., Kim, H. and Riel, M. (1990) Analyzing instructional interactions on electronic message networks, in L. Harasim (ed.) *Online Education: Perspectives in a New Environment*, Praeger, New York, pp. 185–214

Look, H. and Hollar, S. (1999) *Developing a Course Web Page: Virtual Resource Packet*, http://www.knc.lib.umich.edu/guides/CourseWeb/gsihome.html

Luchini, K. (1998) Problems and potentials in web-based instruction, with particular focus on distance learning, *Educational Technology and Society*, **1**(1), 12–13

Lynch, P.J. and Horton, S. (1997) *Yale C/AIM Web Style Guide*, http://info.med.yale.edu/caim/manual/

Madan, V. (1996) Programme evaluation for quality assessment in distance learning, *Quality in Higher Education*, **2**(3), 257–269

Maddux, C. (1996) The state of the art in web-based learning, *Computers in the Schools*, **12**(4), 63–71

Maddux, C., Cummings, R. and Torres-Rivera, E. (1999) Facilitating the integration of information technology into higher-education instruction, *Educational Technology*, **39**(3), 43–47

Malikowski, S. (1997) Interacting in history's largest library: web-based conferencing tools, in B.H. Khan (ed.) *Web-Based Instruction*, Educational Technology Publications, Englewood Cliffs, NJ, pp. 283–298

Martin, P. (1997) Internet for learning explored, *Educational Computing and Technology*, April–May, 26–28

McClure, C. and Lopata, C. (1996) *Assessing the Academic Networked Environment: Strategies and Options*, Association of Research Libraries for the Coalition for Networked Information, Washington, DC

McConigle, D. and Eggers, R.M. (1998) Stages of virtuality: instructor and student, *TechTrends*, **43**(3), 23–26

McCracken, R. (1999) Copyright issues in the production and development of open and distance learning materials in the United Kingdom, *Proceedings of the PanCommonwealth Forum on Open Learning*, Brunei, March 1–5, http://www.col.org/forum/PCFpapers/PostWork/Mccracken.pdf

McVay, M. (1998) Facilitating knowledge construction and communication on the internet, *Technology Source*, December, http://horizon.unc.edu/TS/commmentary/1998-12.asp

Media Matrix (1999) *Press release*, 20 July, New York, http://www.mediamatrix.com

Mirza, J. (1998) Adopting new learning technologies in distance education, *Indian Journal of Open Learning*, **7**(1), 89–94

Mitchell, W.J. (1995) *City of Bits*, MIT Press, Cambridge, MA

Montgomery, A. (1998) Weaving a course based web, *Indiana Higher Education Faculty Papers*, http://www.ihets.org/distance_ed/fdpapers/1998/42.html

Morrison, J.L. (1997a) Using multimedia learning courseware to supplement instruction, *Technology Source*, December, http://horizon.unc.edu/TS/featured/1997-12.asp

Morrison, J.L. (1997b) Higher education and information technology: an interview with Carol Twigg, *Technology Source*, December, http://horizon.unc.edu/TS/commmentary/1997-12.asp

Morrison, J.L. (1998) Technology and higher education: an interview with William Graves, *Technology Source*, January, http://horizon.unc.edu/TS/commmentary/1998-01.asp

National Center for Accessible Media (NCAM) (1999) Captioning and audio description on the web, *Web Access Project*, http://www.wgbh.org/wgbh/pages/ncam/webaccess/index.html (01.09.99)

National Council for Educational Technology (1993) Management of IT and cross-cultural issues, *Directory of Information*, Volume 3, Section 6.37

Naughton, J. (1998) Stripped for action—Playboy versus the pin-up queen, *The Observer Review*, 29 March, p. 9

Negroponte, N. (1995) *Being Digital*, Alfred A. Knopf Inc., New York

Netscape Communications Corporation (1998) *Core JavaScript Reference v1.4*, http://developer.netscape.com/docs/manuals/js/core/jsref/contents.htm (10.29.98)

Newton, R., Marcella, R. and Middleton, I. (1998) NetLearning: creation of an online directory of internet learning resources, *British Journal of Educational Technology*, **29**(2), 173–176

Nichols, G. (1997) Formative evaluation of web-based instruction, in B.H. Khan (ed.) *Web-Based Instruction*, Educational Technology Publications, Englewood Cliffs, NJ, pp. 369–374

Nyiri, J. (1997) Open and distance learning in an historical perspective, *European Journal of Education*, **32**(4), 347–357

O'Carroll, P. (1997) Learning materials on the World Wide Web: text organisation and theories of Learning, *Australian Journal of Adult and Community Education*, **37**(2), 119–123

Orsini-Jones, M. and Davidson, A. (1999) From reflective learners to reflective lecturers via WebCT, *Active Learning*, **10**, 32–38

Osborne, J. (1995) The emergence of convergence: how distance education materials are used in on-campus classes, in D. Stewart (ed.) *One World Many Voices*, Seventeenth World Conference for Distance Education, Volume 2, pp. 167–170

Parker, R.C. (1995) *Desktop Publishing and Design for Dummies*, IDG Books, Foster City, CA

Patel, A. and Hobbs, S. (1998) Does the web offer a solution to many old problems but create new ones in turn?, *Educational Technology and Society*, **1**(1), 10–11

Payne, S.J. (1991) Interface problems and interface resources, in J.M. Carroll (ed.) *Designing Interaction: Psychology at the Human–Computer Interface*, Cambridge University Press, Cambridge, pp. 128–153

Perry, T., Perry, L. and Hosack-Curlin, K. (1998) Internet use by university students: an interdisciplinary study on three campuses, *Internet Research*, **8**(2), 136–141

Peters, G. (1998) From continuing education to lifelong learning: a review of UACE strategy and objectives, Universities Association for Continuing Education Occasional Paper no. 20, http://www.leeds.ac.uk/educol/documents/000000396.htm

Pogue, D. and Schorr, J. (1999) *Macworld Mac Secrets*, fifth edition, IDG Books, Foster City, CA

Pohan, C. and Mathison, C. (1998) WebQuests: the potential of internet based instruction for global education, *Social Studies Review*, **37**(2), 91–93

Polyson, S. (1997) Create an online course in a day, *Journal of Interactive Instruction Development*, **10**(2), 24–27

Potashnik, M. and Capper, J. (1998) Distance education; growth and diversity, *Finance and Development*, **35**(1), 42–45

Pratchett, T. (1994) *Interesting Times*, Corgi, London

Ptaszynski, J.G. (1997a) PowerPoint as a technology enhancement to traditional classroom activities, *Technology Source*, May, http://horizon.unc.edu/TS/featured/1997-05.asp

Ptaszynski, J.G. (1997b) Needed: a beacon, *Technology Source*, March, http://horizon.unc.edu/TS/commmentary/1997-03.asp

Ptaszynski, J.G. (1997c) Technology provides new responses to old problems, *Technology Source*, July, http://horizon.unc.edu/TS/commmentary/1997-07.asp

Ptaszynski, J.G. (1997d) Shared misery/shared solutions: major factors inhibiting the accelerated adoption of technology in higher education, *Technology Source*, April, http://horizon.unc.edu/TS/commmentary/1997-04.asp

Rasmussen, K., Northrup, P. and Lee, R. (1997) Implementing web-based instruction, in B.H. Khan (ed.) *Web-Based Instruction*, Educational Technology Publications, Englewood Cliffs, NJ, pp. 341–346

Ravitz, J. (1997) Evaluating learning networks: a special challenge for web-based instruction, in B.H. Khan (ed.) *Web-Based Instruction*, Educational Technology Publications, Englewood Cliffs, NJ, pp. 361–368

Reed, J. and Afieh, A. (1998) Developing interactive educational engineering software for the World Wide Web with Java, *Computers and Education*, **30**(4), 183–194

Reeves, T.C. and Reeves, P.M. (1997) Effective dimensions of interactive learning on the World Wide Web, in B.H. Khan (ed.) *Web-Based Instruction*, Educational Technology Publications, Englewood Cliffs, NJ, pp. 59–66

Relan, A. and Gillani, B. (1997) Web-based instruction and the traditional classroom: similarities and differences, in B.H. Khan (ed.) *Web-Based Instruction*, Educational Technology Publications, Englewood Cliffs, NJ, pp. 41–46

Retalis, S., Makrakis, V., Papaspyrou, N. and Skordalakis, M. (1998) A case study of an enriched classroom model based on the World Wide Web, *Active Learning*, **8**, 15–19

Reynolds, M. (1997) Learning styles: a critique, *Management Learning*, **28**(2), 115–133

Richards, L. (1997) Distance education: realizing its potential, *International Journal of Engineering Education*, **13**(1), 6–12

Riley, P. (1998) Designing, developing and implementing WWW-based distance learning, *Journal of Interactive Instruction Development*, **10**(4), 18–23

Ritchie, D.C. and Hoffman, B. (1997) Incorporating instructional design principles with the World Wide Web, in B.H. Khan (ed.) *Web-Based Instruction*, Educational Technology Publications, Englewood Cliffs, NJ, pp. 135–138

Rizza, M.G. (1998) Cyberspace and beyond: one view of the possibilities of course web sites, *Indiana Higher Education Faculty Papers*, http://www.ihets.org/distance_ed/fdpapers/1998/43.html

Roberts, D. (1999) Optimising the use of distance learning materials in higher education in Australia, *Proceedings of the PanCommonwealth Forum on Open Learning*, Brunei, March 1–5, http://www.col.org/forum/casestudies.htm

Roberts, D.W. (1998) Using distance education materials for on-campus learning, *Distance Education*, **19**(2), 358–374

Roman, M.B. (1988) When good scientists turn bad, *Discover*, April, 50–58

Romiszowski, A.J. (1997) Web-based distance learning and teaching: revolutionary invention or reaction to necessity?, in B.H. Khan (ed.) *Web-Based Instruction*, Educational Technology Publications, Englewood Cliffs, NJ, pp. 25–37

Rossner, V. and Stockley, D. (1997) Institutional perspectives on organizing and delivering web-based instruction, in B.H. Khan (ed.) *Web-Based Instruction*, Educational Technology Publications, Englewood Cliffs, NJ, pp. 333–336

Rowntree, D. (1998a) Motivating teachers for materials-based learning, *International Journal for Academic Development*, **3**(1), 47–53

Rowntree, D. (1998b) Assessing the quality of materials-based teaching and learning, *Open Learning*, **13**(2), 12–22

Ruppert, S. (1998) Legislative views on higher education technology use, *Thought and Action*, **14**(1), 41–48

Sammons, M.C. (1997) Using Powerpoint presentations in writing classes, *Technology Source*, August, http://horizon.unc.edu/TS/featured/1997-08.asp

Schuttloffel, M.J. (1998) Reflections on the Dilemma of Distance Learning, *International Journal of Educational Telecommunications*, **4**(1), 45–58

Schwier, R. and Misanchuk, E. (1996) Designing multimedia for the hypertext markup language, *Journal of Interactive Instruction Development*, **8**(4), 15–25

Scigliano, J., Levin, J. and Home, G. (1996) Using html for organising student projects through the internet, *THE Journal*, **24**(1), 51–56

Sherry, L. (1996) Issues in distance learning, *International Journal of Educational Telecommunications*, **1**(4), 337–365

Sherry, L. and Wilson, B. (1997) Transformative communication a stimulus to Web innovations, in B.H. Khan (ed.) *Web-Based Instruction*, Educational Technology Publications, Englewood Cliffs, NJ, pp. 67–73

Shiple, J. (1998) Squishy's crash course in information architecture — lesson 3, *Webmonkey*, 15 July, http://www.hotwired.com/webmonkey/98/28/index2a.html

Shipman, F.M., Marshall, C.C., Furuta, R., Brenner, D.A., Hsieh, H.-W. and Kumar, V. (1997) Using networked information to create educational guided paths, *International Journal of Educational Telecommunications*, **3**(4), 383–400

Untangled web: developing teaching on the internet

Shotsberger, P.G. (1997) Emerging roles for instructors and learners in the web-based instruction classroom, in B.H. Khan (ed.) *Web-Based Instruction*, Educational Technology Publications, Englewood Cliffs, NJ, pp. 101–106

Siegel, M.A. and Kirkley, S. (1997) Moving toward the digital learning environment: the future of web-based instruction, in B.H. Khan (ed.) *Web-Based Instruction*, Educational Technology Publications, Englewood Cliffs, NJ, pp. 263–270

Sloane, A. (1997) Learning with the web: experience of using the world wide web in a learning environment, *Computers and Education*, **28**(4), 207–212

Sommer, B. and Anderson, W. (1997) Computer-based lectures using PowerPoint, *Technology Source*, November, http://horizon.unc.edu/TS/featured/1997-11.asp

Sosabowski, A.H., Herson, K. and Lloyd, A.W. (1998) Identifying and overcoming staff resistance to computer based learning and teaching methods, *Active Learning*, **9**, 26–30

Spira, J.L. (1998) Integrating principles of progressive education into technology-based distance learning, *Technology Source*, October, http://horizon.unc.edu/TS/commmentary/1998-10.asp

Stefanov, K., Stoyanov, S. and Nikolov, R. (1998) Design issues of a distance learning course on business on the internet, *Journal of Computer Assisted Learning*, **14**(2), 83–90

Stein, S.D. (1999) *Learning, Teaching and Researching on the Internet: A Practical Guide for Social Scientists*, Longman, Harlow

Strong, W.S. (1994) *The Copyright Book*, fourth edition, MIT Press, Cambridge, MA

Sugrue, B. and Kobus, R.C. (1997) Beyond information: increasing the range of instructional resources on the world wide web, *TechTrends*, **42**(2), 38–42

Tennant, M. (1997) *Psychology and Adult Learning*, Routledge, London

The Nottingham Trent University (1999) *Internet Code of Practice*, http://www.ntu.ac.uk/providing-info/legal/web-cop.html

Thomas, P., Carswell, L., Price, B. and Petre, M. (1998) A holistic approach to supporting distance learning using the internet: transformation, not translation, *British Journal of Educational Technology*, **29**(2), 149–161

Thomas, R. (1998) Net porn alert at work, *The Observer*, 27 September, p. 6

Thyer, B.A., Polk, G. and Gaudin, J.G. (1997) Distance learning in social work education: a preliminary evaluation, *Journal of Social Work Education*, **33**(2), 363–367

Trilling, B. and Hood, P. (1999) Learning technology and education reform in the knowledge age or we're wired, webbed, and windowed, now what?, *Educational Technology*, **39**(3), 5–18

Tuathail, G. and McCormack, D. (1998) Global conflicts on-line: technoliteracy and developing an internet-based conflict archive, *Journal of Geography*, **97**(1), 1–11

University of Minnesota (1998) *Publishing Information on the World Wide Web*, http://fpd.finop.umn.edu/groups/ppd/documents/policy/Publishing_on_WWW.cfm

Usip, E.E. and Bee, R.H. (1998) A discriminant analysis of students' perceptions of web-based learning, *Social Science Computer Review*, **16**(1), 16–29

Van Dusen, G. (1998) Technology: higher education's magic bullet, *Thought and Action*, **14**(1), 59–67

Veash, N. (1999) At the wave of a wand, paper heads for the dustbin of history, *The Observer*, 22 August, p. 5

Veen, J. (1999) Whatever happened to PNG?, *Webmonkey*, 22 February, http://www.hotwired.com/webmonkey/99/09/index0a.html

Voithofer, R.J. (1997) The creation of a web site, in B.H. Khan (ed.) *Web-Based Instruction*, Educational Technology Publications, Englewood Cliffs, NJ, pp. 313–318

Ward, M. and Newlands, D. (1998) Use of the web in undergraduate teaching, *Computers and Education*, **31**(2), 171–184

Watson, J.B. and Rossett, A. (1999) Guiding the independent learner in web-based training, *Educational Technology*, **39**(2), 27–36

Weinstein, P. and Quesada, A. (1997) Education goes the distance, *Technology and Learning*, **17**(8), 23–28

Westell, S. (1999a) An introduction to copyright law, part 2, *Mapping Awareness*, **13**(6), 18–19

Westell, S. (1999b) An introduction to copyright law, part 1, *Mapping Awareness*, **13**(5), 14–15

Whalley, W.B. (1995) Teaching and learning on the internet, *Active Learning*, **2**, 25–29

Wheatley, M. (1996) Catching a cold from the internet, *Human Resources*, July–August, 94

Whittington, C.D. and Campbell, L.M. (1998) Task-oriented learning on the web, *IETI*, **36**(1), 26–33

Williams, D. (1999) Change management: how to really make open learning work, *Proceedings of the PanCommonwealth Forum on Open Learning*, Brunei, March 1–5, http://www.col.org/forum/casestudies

Williams, V. and Peters, K. (1997) Faculty incentives for the preparation of web-based instruction, in B.H. Khan (ed.) *Web-Based Instruction*, Educational Technology Publications, Englewood Cliffs, NJ, pp. 107–110

Willis, B. and Dickinson, J. (1997) Distance education and the World Wide Web, in B.H. Khan (ed.) *Web-Based Instruction*, Educational Technology Publications, Englewood Cliffs, NJ, pp. 81–84

Wilson, B.C. (1999) Evolution of learning technologies: from instructional design to performance support to network systems, *Educational Technology*, **39**(2), 32–35

Windschitl, M. (1998) The WWW and classroom research: what path should we take?, *Educational Researcher*, **27**(1), 28–35

Yaverbaum, G. and Liebowitz, J. (1998) Gofigure Inc: a hyper-media web-based case, *Computers and Education*, **30**(3–4), 147–156

Yeomans, K. (1998) *Learners on the Superhighway? Access to Learning via Electronic Communication*, Winston Churchill Fellowship Report, National Institute of Adult Continuing Education, Leicester

Yong, Y. and Wang, S. (1996) Faculty perceptions on a new approach to distance learning—TELETECHNET, *Journal of Instruction Delivery Systems*, **10**(2), 3–5

Zellner, R.D. (1997) Management of instructional materials and student products in web-based instruction, in B.H. Khan (ed.) *Web-Based Instruction*, Educational Technology Publications, Englewood Cliffs, NJ, pp. 347–352

Zepke, N. (1998) Instructional design for distance delivery using hypertext and the internet: assumptions and applications, *Quality in Higher Education*, **4**(2), 173–186

Zobel, S.M. (1997) Legal implications of intellectual property and the World Wide Web, in B.H. Khan (ed.) *Web-Based Instruction*, Educational Technology Publications, Englewood Cliffs, NJ, pp. 337–340

index

index

Note: Page references in *italics* refer to Figures; those in **bold** refer to Tables